Lecture Notes in Physics

Volume 958

The Lecture Notes in Physics

The series Lecture Notes in Physics (LNP), founded in 1969, reports new developments in physics research and teaching-quickly and informally, but with a high quality and the explicit aim to summarize and communicate current knowledge in an accessible way. Books published in this series are conceived as bridging material between advanced graduate textbooks and the forefront of research and to serve three purposes:

- to be a compact and modern up-to-date source of reference on a well-defined topic
- to serve as an accessible introduction to the field to postgraduate students and nonspecialist researchers from related areas
- to be a source of advanced teaching material for specialized seminars, courses and schools

Both monographs and multi-author volumes will be considered for publication. Edited volumes should, however, consist of a very limited number of contributions only. Proceedings will not be considered for LNP.

Volumes published in LNP are disseminated both in print and in electronic formats, the electronic archive being available at springerlink.com. The series content is indexed, abstracted and referenced by many abstracting and information services, bibliographic networks, subscription agencies, library networks, and consortia.

Proposals should be sent to a member of the Editorial Board, or directly to the managing editor at Springer:

Lisa Scalone
Springer Nature
Physics Editorial Department
Tiergartenstrasse 17
69121 Heidelberg, Germany
Lisa.Scalone@springernature.com

More information about this series at http://www.springer.com/series/5304

Simone Marzani • Gregory Soyez •
Michael Spannowsky

Looking Inside Jets

An Introduction to Jet Substructure and Boosted-object Phenomenology

 Springer

Simone Marzani
Dipartimento di Fisica
Università di Genova
Genova, Italy

Gregory Soyez
Institut de Physique Theorique
CNRS UMR 3681, CEA Saclay
Gif-sur-Yvette cedex, France

Michael Spannowsky
Department of Physics
Institute for Particle Physics
Phenomenology
Durham University
Durham, UK

ISSN 0075-8450   ISSN 1616-6361   (electronic)
Lecture Notes in Physics
ISBN 978-3-030-15708-1   ISBN 978-3-030-15709-8   (eBook)
https://doi.org/10.1007/978-3-030-15709-8

Library of Congress Control Number: 2019936176

This Springer imprint is published by the registered company Springer Nature Switzerland AG.
The registered company address is: Gewerbestrasse 11, 6330 Cham, Switzerland

Preface

The study of the internal structure of hadronic jets has become in recent years a very active area of research in particle physics. Jet substructure techniques are increasingly used in experimental analyses by the Large Hadron Collider collaborations, in the context of searching both for new physics and for Standard Model measurements. On the theory side, the quest for a deeper understanding of jet substructure algorithms has contributed to a renewed interest in all-order calculations in Quantum Chromodynamics (QCD). This has resulted in new ideas about how to design better observables and how to provide a solid theoretical description for them. In the last few years, jet substructure has seen its scope extended, for example, with an increasing impact in the study of heavy-ion collisions, or with the exploration of deep-learning techniques. Furthermore, jet physics is an area in which experimental and theoretical approaches meet together, where cross-pollination and collaboration between the two communities often bear the fruits of innovative techniques. The vivacity of the field is testified, for instance, by the very successful series of BOOST conferences together with their workshop reports, which constitute a valuable picture of the status of the field at any given time.

However, despite the wealth of literature on this topic, we feel that a comprehensive and, at the same time, pedagogical introduction to jet substructure is still missing. This makes the endeavour of approaching the field particularly hard, as newcomers have to digest an increasing number of substructure algorithms and techniques, too often characterised by opaque terminology and jargon. Furthermore, while first-principle calculations in QCD have successfully been applied in order to understand and characterise the substructure of jets, they often make use of calculational techniques, such as resummation, which are not the usual textbook material. This seeded the idea of combining our experience in different aspects of jet substructure phenomenology to put together this set of lecture notes, which we hope could help and guide someone who moves their first steps in the physics of jet substructure.

Genova, Italy   Simone Marzani
Gif-sur-Yvette cedex, France   Gregory Soyez
Durham, UK   Michael Spannowsky

Acknowledgements

Most of (if not all) the material collected in this book comes from years of collaboration and discussions with excellent colleagues who helped us and influenced us tremendously. In strict alphabetical order, we wish to thank Jon Butterworth, Matteo Cacciari, Mrinal Dasgupta, Frederic Dreyer, Danilo Ferreira de Lima, Steve Ellis, Deepak Kar, Roman Kogler, Phil Harris, Andrew Larkoski, Peter Loch, David Miller, Ian Moult, Ben Nachman, Tilman Plehn, Sal Rappoccio, Gavin Salam, Lais Schunk, Dave Soper, Michihisa Takeuchi, Jesse Thaler, and Nhan Tran. We would also like to thank Frederic Dreyer, Andrew Lifson, Ben Nachman, Davide Napoletano, Gavin Salam, and Jesse Thaler for their helpful suggestions and comments on the manuscript.

Contents

Chapter 1
Introduction and Motivation

The Large Hadron Collider (LHC) at CERN is the largest and most sophisticated machine to study the elementary building blocks of nature ever built. At the LHC protons are brought into collision with a large centre-of-mass energy—7 and 8 TeV for Run I (2010–13), 13 TeV for Run II (2015–18) and 14 TeV from Run III (starting in 2021) onwards—to resolve the smallest structures in a controlled and reproducible environment. As protons are not elementary particles themselves, but rather consist of quarks and gluons, their interactions result in highly complex scattering processes, often with final state populated with hundreds of particles, which are measured via their interactions with particle detectors.

Jets are collimated sprays of hadrons, ubiquitous in collider experiments, usually associated with the production of an elementary particle that carries colour charge, e.g. quarks and gluons. Their evolution is governed by the strong force, which within the Standard Model of particle physics is described by Quantum Chromodynamics (QCD). The parton (i.e. quark or gluon) that initiates a jet may radiate further partons and produce a (collimated) shower of quarks and gluons, a so-called parton shower, that eventually turn into the hadrons (π, K, p, n, ...) observed in the detector. The vast majority of LHC events (that one is interested in) contain jets. They are the most frequently produced and most complex objects measured at the LHC multipurpose experiments, ATLAS and CMS.

When protons collide inelastically with a large energy transfer between them, one can formally isolate a hard process at the core of the collision, which involves one highly-energetic parton from each of the two protons. These two partons interact and produce a few elementary particles, like two partons, a Higgs boson associated with a gluon, a top–anti-top pair, new particles, ... Since the energy of this hard process is large, typically between 100 GeV and several TeV, there is a large gap between the incoming proton scale and the hard process on one hand, and between the hard process and the hadron scale on the other. This leaves a large phase-space for parton showers to develop both in the initial and final state of the collision. This picture is clearly a simplification because we can imagine that secondary parton-

© Springer Nature Switzerland AG 2019
S. Marzani et al., *Looking Inside Jets*, Lecture Notes in Physics 958,
https://doi.org/10.1007/978-3-030-15709-8_1

parton interactions might take place. These multi-parton interactions constitute what is usually referred to as the *Underlying Event*. To complicate things further, the LHC does not collide individual protons, but bunches of $\mathcal{O}(10^{11})$ protons. During one bunch crossing it is very likely that several of the protons scatter off each other. While only one proton pair might result in an event interesting enough to trigger the storage of the event on tape, other proton pairs typically interact to give rise to hadronic activity in the detectors. This additional hadronic activity from multiple proton interactions is called *pileup*. On average, radiation from pileup is much softer than the jets produced from the hard interaction, but for jet (and jet substructure) studies it can have a significant impact by distorting the kinematic relation of the jet with the hard process.

In recent years the detailed study of the internal structure of jets has gained a lot of attention. At LHC collision energy electroweak (EW) scale resonances, such as the top quark, W/Z bosons and the Higgs boson, are frequently produced beyond threshold, i.e. their energy (transverse momentum) can significantly exceed their mass. Therefore, analyses and searching strategies developed for earlier colliders, in which EW-scale particles were produced with small velocity, have to be fundamentally reconsidered. Because EW resonances decay dominantly into quarks, when they are boosted, their decay products can become collimated in the lab-frame and result in one large and massive jet, often referred to as a fat jet. Initially such a configuration was considered disadvantageous in separating processes of interest (i.e. processes which included EW resonances) from the large QCD backgrounds (where jets are abundantly produced from high-energy quarks and gluons). However, with the popularisation of sequential jet clustering algorithms retaining the full information of the jet's recombination history, it transpired that one can use the internal structure of jets to tell apart jets that were induced by a decaying boosted EW resonance or by a QCD parton. *This investigation of the internal structure of jets is what one refers to as jet substructure.*

While the first jet substructure methods have been put forward in the 1990s and early 2000s [1–4], it was only in 2008, with the proposal to reconstruct the Higgs boson in vector-boson associated production [5], that the interest in understanding and utilising jet substructure surged tremendously [6–11]. If the Higgs boson, being spin and colour-less, the perfect prototype of a featureless resonance could be reconstructed, surely other EW-scale resonances proposed in many extensions of the Standard Model could be discovered as well. Furthermore, jet substructure can be exploited in searches of physics beyond the Standard Model (BSM) not necessarily restricted to the EW scale. For instance, in many such extensions TeV-scale resonances are predicted which decay subsequently into EW particles, which could either be Standard Model or BSM resonances. Because of the mass differences, these EW-particles are typically boosted and their hadronic decay might be reconstructed as a fat jet. Thus, scenarios where jet substructure methods can benefit searches for BSM physics are rather frequent.

Boosted Resonances in New Physics Searches

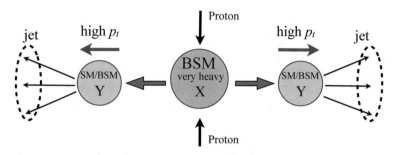

Fig. 1.1 Generic interaction sequence for the search of a very BSM resonance that decays into electroweak-scale particles that subsequently decay hadronically

A typical situation of interest for BSM searches using jet substructure is illustrated in Fig. 1.1. A heavy new resonance X with a mass of $\mathcal{O}(1)$ TeV is produced in a proton-proton collision. This heavy BSM resonance quickly decays into lighter states Y—e.g. W/Z/H bosons or lighter BSM particles—with a mass around the EW scale. Particles Y are typically produced with large transverse momentum (p_t) because their mass is much smaller than the mass of the decaying particle X. Finally, if a particle Y decays hadronically, because of its large boost, its decay product in the lab frame are collimated and reconstructed into a jet. The aim of jet substructure is therefore to distinguish a *signal jet*, originated from a boosted massive particles, such as Y, from *background jets*, which typically are QCD jets originated from quarks and gluons.

Consequently various ways of discriminating the sources of jets have been devised, with the aim to classify a jet as of interest for a specific search or measurement or not. Most methods to achieve this classification task follow a two-step approach: firstly, the jet is cleaned up (*groomed*), i.e. soft radiation which is unlikely to come from the decaying resonance is removed, and, secondly, one computes observables specifically designed to separate signal and background jets based on the energy distribution amongst the remaining jet constituents. Step two could be subdivided further into two classes of classifiers: *jet-shape observables* and *prong-finders*. Jet-shape observables only consider the way the energy is spatially distributed inside a jet, e.g. they do not take into account the recombination history of the fat jet itself. Prong-finders instead aim to construct hard subjets inside a fat jet, i.e. isolated islands of energy inside the jet, and compare properties of subjets, potentially including information on their formation in the fat jet's recombination history.

Both jet shapes and prong finders aim to disentangle the different topologies that characterise signal and background jets. For instance, QCD jets are characterised by a hard core surrounded by soft/collinear radiation, leading predominantly to jets with a one-prong structure. EW bosons instead, such as W/Z and the Higgs, decays in a quark-antiquark pair, which roughly share equal fractions of the heavy particle momentum, leading to a two-prong structure. Finally, the top quark preferentially decays into a bottom quark and a W boson, which then decays in a pair of light quarks. Hence, top-initiated jets features a three-pronged structure. It has been shown that grooming techniques and jet substructure observables are sensitive to different effects during the complex evolution of a jet, hence the classification of jets benefits from combining various of these techniques [9, 12].[1] Thus, by combining groomers and different subjet observables, high-level tagging methods can be constructed for the reconstruction of top quarks, W/Z and Higgs bosons and new-physics resonances.

Nowadays, the application of jet substructure techniques has considerably widen and goes well beyond the identification of massive boosted particles. A specific example particularly relevant for this book is that because grooming techniques reduce an observable's sensitivity to soft physics, comparisons between experimental data and first-principle calculations are less affected by non-perturbative contamination. Consequently, the catalogue of Standard Model measurements with jet substructure techniques keeps growing. Furthermore, jet substructure techniques have found applications also in initially unexpected ways. For instance, it has been realised that substructure variables can be used to probe the jet interaction with the quark-gluon plasma in heavy-ion collisions, providing new observables helping to improve our understanding of this difficult question. Finally, particle physics in general, and jet physics in particular, is enjoying a period of rapid development as innovative ideas and techniques exploiting machine-learning are poured into the field. Unfortunately, this topic goes beyond the scope of this book and we refer the interested reader to the recent review [10].

Although this book focuses on LHC physics, it is worth pointing out that jet substructure techniques have also been used at other colliders, such as the Tevatron or RHIC. Due to the lower collision energy, the scope of substructure studies is more limited. We can however point the readers to Refs. [7, 13] for reviews of substructure studies at the Tevatron and to Ref. [14] for an explicit measurement by the STAR collaboration at RHIC.

These lecture notes aim to provide an accessible entry—at the level of graduate students with some expertise in collider phenomenology—to the quickly growing field of jet substructure physics. Due to the complexity of the internal structure of jets, this topic connects to subtle experimental and quantum-field theoretical questions. In order to make these notes as self-contained as possible, the first four chapters will provide a broad introduction to jet physics and related QCD ingredients. First, we will give a brief introduction into QCD and its application to collider phenomenology in Chap. 2, focusing on those aspects that are needed the most in jet physics. Chapter 3 will introduce the basics of jet definition and

---

[1]Finding hard subjets (the task of prong-finders) and removing soft contamination (the task of groomers) are similar in practice. This means that tools which do one, very often do the other as well.

jet algorithms, including some of the experimental issues related to defining and measuring jets. In Chap. 4, we will discuss in some detail a key observable in jet physics, namely the jet invariant mass. We will show how its theoretical description requires an all-order perturbative approach and we will discuss various aspects of this resummation. We will dive into the topic of modern jet substructure in Chap. 5 where we will first describe the main concepts and ideas behind substructure tools and then try to give an comprehensive list of the different approaches and tools which are currently employed by the substructure community (theoretical and experimental). Chapters 6–9 explore our current first-principle understanding of jet substructure with each chapter addressing a different application. First, in Chap. 6 we discuss groomers which have been the first tools for which an analytic understanding became available. In particular, we will go back to the jet mass and we will study in detail how its distribution is modified if grooming techniques are applied. In the remaining chapters, we will discuss more advanced topics such as quark/gluon discrimination in Chap. 7, two-prong taggers in Chap. 8 and, finally, Sudakov safety in Chap. 9. Finally, in the last part of this book, we will discuss the current status of searches and measurements using jet substructure in Chap. 10.

A large part of these lecture notes will focus on our current first-principle understanding of jet substructure in QCD. The key observation to keep in mind in this context is the fact that substructure techniques are primarily dealing with boosted jets, for which the transverse momentum, p_t, is much larger than the mass, m. From a perturbative QCD viewpoint, this means that powers of the strong coupling will be accompanied with large logarithms of p_t/m, a common feature of QCD whenever we have two largely disparate scales. For these situations, a fixed-order perturbative approach is not suited and one should instead use all-order, *resummed*, calculations which focus on including the dominant logarithmically-enhanced contributions at all orders in the strong coupling. Chapter 4 will present a basic introduction to resummation taking the calculation of the jet mass as a practical example.

There exist different approaches on how to tackle this type of calculations. On the one hand, one could analyse the structure of matrix elements for an arbitrary number of quark and gluon emissions in the soft/collinear limit and from that derive the all-order behaviour of the distribution of interest. In this context, the coherent branching algorithm [15, 16] deserves a special mention because not only it is the basis of angular-ordered parton showers, but it also constitutes the foundation of many resummed calculations (for a review see e.g. [17]). Other approaches to all-order resummation instead take a more formal viewpoint and try to establish a factorisation theorem for the observable at hand, therefore separating out the contribution from hard, soft and collinear modes. This point of view is for instance, the one taken when calculations are performed in Soft-Collinear Effective Theory (SCET). For a pedagogical introduction to SCET, we recommend Ref. [18].

In this book, we will use the former approach, but we will try to point out the relevant literature for SCET-based calculations too. That said, our aim is not to present a rigorous and formal proof of resummed calculations, but rather to lay out the essential ingredients that go into these theoretical predictions, while keeping the discussion at a level which we think it is understandable for readers with both theoretical and experimental backgrounds. In particular, even though Chaps. 6–9 start with (sometimes heavy) analytic QCD calculations, we will always come back to comparisons between these analytic calculations and Monte Carlo simulations in the end. This will allow us to discuss the main physical features of the observed distributions and how they emerge from the analytic understanding. It will also allow us to discuss how the analytic results obtained in perturbative QCD are affected by non-perturbative corrections.

Chapter 2
Introduction to QCD at Colliders

Jet physics is QCD physics. Therefore, a solid and insightful description of jets and their substructure relies on a deep understanding of the dynamics of strong interactions in collider experiments. QCD is an incredibly rich but, at the same time, rather complicated theory and building up a profound knowledge of its workings goes beyond the scope of this book. At the same time, some familiarity with perturbative calculations in quantum field theory is necessary in order to proceed with our discussion. Therefore, in this chapter we recall the essential features of the theory of strong interactions that are needed in jet physics. Because we aim to make this book accessible to both theorists and experimenters that want to move their first steps in jet substructure, we are going to take a rather phenomenological approach and we will try to supplement the lack of theoretical rigour with physical intuition. QCD itself helps us in this endeavour because the dynamics that characterises jet physics is often dominated by soft and collinear radiation, i.e. emissions of partons that only carry a small fraction of the hard process energy or that are emitted at small angular distances. The structure of the theory greatly simplifies in this limit and many results can be interpreted using semi-classical arguments. The price we have to pay is that, if we want to achieve a reliable description of observables in the soft and collinear regions of phase-space, we have to go beyond standard perturbation theory and consider the summation of some contributions to all orders in perturbative expansion.

2.1 The Theory of Strong Interactions

Let us begin our discussion with a historical detour. The quest for a coherent description of strong interactions started in the 1960s and had the principal aim of understanding and classifying the plethora of new particles produced at the first particle colliders. Indeed, as machines to accelerate and collide particles were

© Springer Nature Switzerland AG 2019
S. Marzani et al., *Looking Inside Jets*, Lecture Notes in Physics 958,
https://doi.org/10.1007/978-3-030-15709-8_2

becoming more powerful, many new strongly-interacting particles, collectively referred to as hadrons, were produced, leading to what was defined as a particle zoo. Some of these particles shared many similarities to the well-known protons, neutrons and pions and could therefore be interpreted as excited states of the formers. Other particles instead presented new and intriguing properties. A major breakthrough was realised with the *quark model*. This model successfully applied the formalism of group theory to describe the quantum numbers of the hadrons known at that time. It introduced fundamental constituents with fractional electric charge called quarks and described mesons and baryons in terms of the different combinations of these constituents. However, the model made no attempt to describe the dynamics of these constituents. The quark model led to another important discovery: the introduction of a new degree of freedom, which was termed colour. Its introduction was made necessary in order to recover the symmetry properties of the wave-function of some baryonic states such as the Δ^{++} or the Ω^-.

Alongside hadron spectroscopy, scattering processes were used to study the structure of the hadrons. In this context, experiments where beams of electrons were scattered off protons played a particular important role, as they were used to probe the structure of the protons at increasingly short distances. The experiments in the deep-inelastic regime, where the target protons were destroyed by the high-momentum-transfer interaction with the electron, pointed to peculiar results. The interaction was not between the electron and the proton as a whole, but rather with pointlike constituents of the proton, which behaved as almost-free particles. In order to explain these experimental data, the *parton model* was introduce in the late Sixties. The basic assumption of this model is that in high-energy interactions, hadrons behave as made up of almost free constituents, the partons, which carry a fraction of the hadron momentum. Thus, the description of the hadron is given in terms of partonic distributions that represent the probability of having a particular parton which carries a fraction of the total hadron's momentum.

The quark model and the parton model aim to describe rather different physics: the former classifies the possible states of hadronic matter, while the latter applies if we want to describe how a hadron interacts at high energy. However, it is very suggestive that they both describe hadronic matter as made up of more elementary constituents. A successful theory of the strong force should be able to accommodate both models. Nowadays Quantum Chromo-Dynamics (QCD) is accepted as the theory of strong interactions. It is a non-Abelian gauge theory and the symmetry group is the local version of the colour symmetry group SU(3). The theory describes the interaction between fermionic and bosonic fields associated to quarks and gluons respectively (see for instance [19–23] and references therein).

The QCD Lagrangian

$$\mathcal{L} = -\frac{1}{4} F_{\mu\nu}^A F_A^{\mu\nu} + \sum_{\text{flavours}} \bar{\psi}_a (i\gamma_\mu D^\mu - m)_{ab} \psi_b \,, \tag{2.1}$$

where $F_{\mu\nu}^A$ is the gluon field strength, defined by:

$$F_{\mu\nu}^A = \partial_\mu A_\nu^A - \partial_\nu A_\mu^A + g_s f^{ABC} A_\mu^B A_\nu^C \,. \tag{2.2}$$

and D_μ is the covariant derivative

$$\left(D_\mu\right)_{ab} = \partial_\mu \delta_{ab} - i g_s A_\mu^A t_{ab}^A, \tag{2.3}$$

where t^A are the algebra generators. In the above equations both lower-case and upper-case indices indicate refer to SU(3), the formers denote indices in the (anti-)fundamental representation, while the latter in the adjoint one. We note that a sum over quark flavours, namely up, down, charm, strange, top, and bottom is indicated. Strong interactions are completely blind to this quantum number and therefore the only distinction between different quark flavours in this context comes about only because of the mass. Note that the quark masses span several orders of magnitude and therefore the related phenomenology is extremely different!

A remarkable feature of QCD is the fact that the strong coupling $\alpha_s = g_s^2/4\pi$ is a decreasing function of the energy involved in the process. For this reason QCD has a low energy regime, in which the theory is strongly-interacting and a high-energy one, in which it is asymptotically free. This implies that strong processes are computable in perturbation theory if a sufficiently high-energy scale is involved. Thus, asymptotic freedom provides the theoretical justification of the parton model, which can be understood as the lowest order approximation of a perturbative QCD calculation.

The theoretical description of high energy collisions of protons is fairly complex. In a typical event hundreds of particles are produced, as depicted in Fig. 2.1. The short-distance, i.e. high-energy, part of the process can be computed using perturbation theory, however long-distance physics is driven by the non-perturbative nature of QCD at low energy scales. Fortunately, there exists a theorem in QCD that enables us to separate the perturbative, i.e. calculable, part of a process from the non-perturbative one, which can be described in terms of parton distribution (or fragmentation) functions. These objects essentially generalise the probability distributions introduced by the parton model. Parton distributions are universal, i.e. they do not depend on the particular process, and they can be determined by fitting data from previous experiments. This is the collinear factorisation theorem and although it has been explicitly proven only for a few processes (deep inelastic scattering of an electron off a proton and the Drell-Yan process), it is usually considered valid and is used ubiquitously in perturbative QCD calculations.[1] In

---

[1] However, examples of short-distance processes that exhibits collinear factorisation breaking have been identified and studied [24–27].

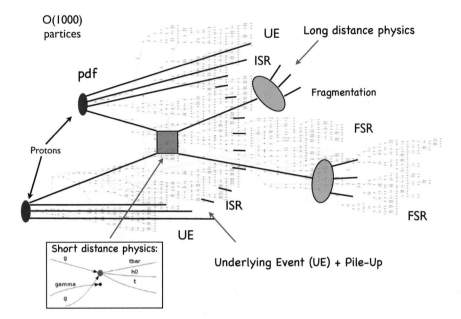

Fig. 2.1 A schematic representation of a typical high-energy proton-proton collision

collinear factorisation, the total cross section of inelastic proton-proton scattering
to produce a final state n can be calculated with the formula

$$\sigma = \sum_{a,b} \int_0^1 dx_a dx_b \int d\Phi_n f_a^{h_1}(x_a, \mu_F) f_b^{h_2}(x_b, \mu_F) \frac{1}{2\hat{s}} |\mathcal{M}_{ab\to n}|^2 (\Phi_n; \mu_F, \mu_R) ,$$
(2.4)

where $f_a^h(x, \mu)$ denotes the parton distribution functions, which depend on the
longitudinal momentum fraction x of parton a with respect to its parent hadron h,
and on an arbitrary energy scale called factorisation scale μ_F. In the above equation,
$d\Phi_n$ denotes the differential phase space element over n final-state particles,

$$d\Phi_n = \prod_{i=1}^n \frac{d^3 p_i}{(2\pi)^3 2E_i} (2\pi)^4 \delta^{(4)}(p_a + p_b - \sum_{i=1}^n p_i) ,$$
(2.5)

where p_a are p_b are the initial-state momenta. The convolution of the squared
matrix element $|\mathcal{M}_{ab\to n}|^2$, averaged over initial-state spin and colour degrees of
freedom, with the Lorentz-invariant phase space Φ_n and multiplied by the flux
factor $1/(2\hat{s}) = 1/(2x_a x_b s)$ results in the calculation of the parton-level cross
section $\hat{\sigma}_{ab\to n}$. The *cross section master formula* of Eq. (2.4) holds to all orders in

perturbation theory, up to terms which are suppressed by $\left(\frac{\Lambda_{QCD}^2}{Q_{min}^2}\right)^p$, where Λ_{QCD} is the non-perturbative QCD scale, Q_{min} is the minimum hard energy scale probed by the process, and typically $p = 1$. For instance, in the case of the inclusive jet cross-section, we typically have $Q_{min} = p_t$, the jet transverse momentum. In what follows we will spend plenty of time discussing the invariant mass m of a jet with large transverse momentum p_t. In that case, we will be able to identify $Q_{min} = m$.

Protons consist of many partons, each carrying a fraction of the total proton energy. The partons of the two protons that interact with each other via a large momentum transfer and the wide gap between this hard scale the proton mass scale is typically filled by the emission of extra partons, which is usually referred to as *initial state radiation*. Furthermore, because the hard momentum transfer can be much smaller than the proton collision energy (13 TeV), initial-state radiation is not necessarily soft. In the hard process, large interaction scales and momentum transfers are probed. New heavy particles can be produced and novel interactions can be tested. Thus, the nature of the hard interaction process leaves a strong imprint in the topological structure and the composition of the whole final state. However, if colour-charged particles[2] are produced during the hard interaction process, they are likely to emit further partons, i.e. *final state radiation*, to evolve from the hard interaction scale down to the hadronisation scale $\mathcal{O}(\Lambda_{QCD})$, where non-perturbative processes rearrange the partons into colour-neutral hadrons.

The proton's energy carried by the spectator partons, i.e. partons of the proton that are not considered initial states of the hard interaction process, is mostly directed into the forward direction of the detector, but a non-negligible amount of radiation off these spectator partons can still end up in the central region of the detector. This so-called Underlying Event (UE), contributing to the measured radiation in a detector, is, on average, softer, i.e. has lower transverse momentum, than for example the decay products of the hard process or initial state radiation. For jet substructure observables, however, it plays an important role as it can complicate the extraction of information from observables that rely on the details of the energy distribution inside a jet.

Furthermore, protons are accelerated and collided in bunches. When two bunches of protons cross at an interaction point, multiple proton-proton collisions can occur simultaneously. What is observed in the detectors is therefore a superposition of these many events. When one of these collisions is hard and deemed interesting enough by the experiments' triggers to be stored on tape, it therefore overlays in the detector with all the other simultaneous, mostly soft, collisions. This effect is known as pileup and presents a challenge to the reconstruction of the objects seen in the detectors in general and of the hadronic part of the event, in particular. To give

---

[2]We are focusing here on QCD-induced parton showers. EW interactions can also give rise to parton showers, however, due to $\alpha \ll \alpha_s$ their contributions are suppressed. However, it should be noted the impact of EW corrections increases with the energy and so it becomes imperative to consistently include them in order to perform accurate phenomenology at future higher-energy colliders.

a quantitative estimate, at the end of Run II of the LHC (late 2018), the machine delivers a luminosity $\mathcal{L} \sim 2 \times 10^{34}\,\text{cm}^{-2}\,\text{s}^{-1}$ which, for a bunch spacing of 25 ns and a typical total proton-proton cross-section of 100 mb, corresponds to an average of 50 interactions per bunch-crossing (assuming that they are Poisson-distributed). We refer the interested reader to a recent review on this subject in the context of jet physics, written by one of us [28].

2.2   Generalities on Perturbative Calculations

The calculation of the matrix element in Eq. (2.4) is usually approximated by a perturbative series in powers of the strong coupling, henceforth the *fixed-order* expansion. The evaluation of such perturbative expansion, and more generally the development of improved techniques to compute *amplitudes*, is one of the core activities of QCD phenomenology. In this framework, theoretical precision is achieved by computing cross-sections σ including increasingly higher-order corrections in the strong coupling α_s

$$\sigma(v) = \sigma_0 + \alpha_s\,\sigma_1 + \alpha_s^2\,\sigma_2 + \alpha_s^3\,\sigma_3 + \mathcal{O}(\alpha_s^4), \tag{2.6}$$

where v is a generic observable, which for definiteness we take dimensionless. In the above expression leading order (LO) contribution σ_0 is the Born-level cross section for the scattering process of interest. Subsequent contributions in the perturbative expansion σ_i constitute the next-to-i-leading (N^iLO) corrections. In the language of Feynman diagrams, each power of α_s corresponds to the emission of a QCD parton, either a quark or a gluon, in the final state or to a virtual correction. The theoretical community has put a huge effort in computing higher-order corrections. LO cross-sections can be computed for an essentially arbitrary number of external particles. Automation has been achieved in recent years also for NLO calculations and an increasing number of NNLO calculations is now available in computer programs. Moreover, for hadron-collider processes with simple topologies, recent milestone calculations have achieved N^3LO accuracy [29, 30]. A particularly important example which falls under this category is the main production channel of the Higgs boson (through gluon-gluon fusion). One of the main challenges in this enterprise is the treatment of the infra-red region. As it is going to be discussed in the following, the emissions of soft and/or collinear partons is also problematic because it can generate large logarithmic terms in the perturbative coefficients, thus invalidating the fixed-order approach.

It is well known that the calculations of Feynman diagrams is plagued by the appearance of divergences of different nature. Loop-diagrams can exhibit ultra-violet singularities. Because QCD is a renormalisable theory, such infinities can be absorbed into a redefinition of the parameters that enter the Lagrangian, e.g. the strong coupling α_s. Moreover, real-emission diagrams exhibit singularities in particular corners of the phase-space. More specifically, the singular contributions have to do with collinear, i.e. small angle, splittings of massless partons and emissions of soft gluons, off both massless and massive particles. Virtual diagrams also exhibit analogous infra-red and collinear (IRC) singularities and rather general theorems [31–33] state that such infinities cancel at each order of the perturbative series Eq. (2.6), when real and virtual corrections are added together, thus leading to observable transition probabilities that are free of IRC singularities. We will explicitly discuss infra-red singularities in a NLO calculation in the next section. Moreover, in order to be able to use the perturbative expansion of Eq. (2.6), one has to consider observables v that are infra-red and collinear (IRC) safe, i.e. measurable quantities that do not spoil the above theorems. We will come back to a more precise definition of IRC safety in Sect. 2.4.

It is worth pointing out that, in practice, non-perturbative effects like hadronisation regulate soft and collinear divergences, so that cross-sections are finite. The requirement of IRC safety means that an observable can be computed reliably in perturbative QCD, up to non-perturbative power corrections, which decrease as the hard scale of the process increases. Moreover, from an experimental viewpoint, the finite resolution of the detectors also acts as a regulator, thus preventing the occurrence of actual singularities. However, this in turn would be reflected on a possibly strong dependence of theoretical predictions on the detector resolution parameters, which one wishes to avoid.

The fixed-order expansion of Eq. (2.6) works well if the measured value of the observable is $v \simeq 1$, a situation in which there is no significant hierarchy of scales. However, it loses its predictive power if the measurement of $v \ll 1$ confines the real radiation into a small corner of phase-space, while clearly leaving virtual corrections UE-restricted. For IRC safe observables the singular terms still cancel, but logarithmic corrections in v are left behind, causing the coefficients σ_i to become large, so that $\alpha_s^i \sigma_i \sim 1$. Because these logarithmic corrections are related to soft and/or collinear emissions, one can expect at most two powers of $L = \log\left(\frac{1}{v}\right)^3$ for each power of the strong coupling. For example, when v is sensitive only to angles up to $\theta_{\text{cut}} \ll 1$, one should expect large (collinear) logarithms of $1/\theta_{\text{cut}}$, and when v is sensitive only to $|k_{3\perp}|$ up to $|k_{3\perp}^{\text{cut}}| \ll 1$, one should expect large (soft) logarithms of $Q/|k_{3\perp}^{\text{cut}}|$.

Let us consider the *cumulative* cross-section for measuring a value of the observable of interest which is less than a given value v, normalised to the inclusive Born-level cross-section σ_0.[4] We have

$$\Sigma(v) = \int_0^v dv' \frac{1}{\sigma_0} \frac{d\sigma}{dv'} \tag{2.7}$$

$$= 1 + \alpha_s \left(\sigma_{12} L^2 + \sigma_{11} L + \dots\right) + \alpha_s^2 \left(\sigma_{24} L^4 + \sigma_{23} L^3 + \dots\right)$$

$$+ \mathcal{O}(\alpha_s^n L^{2n}). \tag{2.8}$$

---

[3]Throughout this book we denote with $\log(x)$ the natural logarithm of x.

[4]Note that in the literature, Σ sometimes refers to the un-normalised cumulative cross-section.

All-order resummation is then a re-organisation of the above perturbative series. For many observables of interest, the resummed expression exponentiates, leading to

$$\sigma(v) = \sigma_0 \, g_0 \exp\left[L g_1(\alpha_s L) + g_2(\alpha_s L) + \alpha_s g_3(\alpha_s L) + \ldots\right], \tag{2.9}$$

where g_0 is a constant contribution which admits an expansion in α_s. In analogy to the fixed-order terminology, the inclusion of the contribution g_{i+1}, $i \geq 0$, leads to next-to^i-leading logarithmic ($N^i LL$) accuracy.

Fixed-order Eq. (2.6) and resummed Eq. (2.9) expansions are complementary. On the one hand, fixed-order calculations fail in particular limits of phase-space, indicating the need for an all-order approach. On the other hand, all-order calculations are only possible if particular assumptions on the emission kinematics are made. Thus, the most accurate theoretical description for the observable v is achieved by matching the two approaches e.g. using (other so-called *matching schemes* exist)

$$\sigma^{\text{matched}}(v) = \sigma^{\text{fixed-order}}(v) + \sigma^{\text{resummed}}(v) - \sigma^{\text{double counting}}(v). \tag{2.10}$$

2.3 Factorisation in the Soft and Collinear Limits

In order to highlight the structure of IRC singularities in matrix elements, we consider the calculation of the NLO QCD corrections in the soft limit. For this presentation we closely follow the review [17]. In order to simplify our discussion, rather than presenting a calculation for proton-proton collisions, for which we would have to include parton distribution functions and discuss how to treat initial-state radiation, we focus our discussion on a process in electron-positron collisions, for which we can concentrate on QCD radiation off the final-state quarks. We will show that the requirement of IRC safety implies with some constraints on observables to guarantee the cancellation of divergences when combining real and virtual diagrams. Furthermore, we will also see that, if we consider an inclusive observable, we obtain an NLO correction which is free of large logarithms.

Let us therefore consider the $\mathcal{O}(\alpha_s)$ correction to the process

$$e^+ e^- \to \gamma^* \to q\bar{q}. \tag{2.11}$$

The relevant Feynman diagrams are shown in Fig. 2.2, where for convenience we have dropped the initial-state lepton line. We label the momentum of the quark and anti-quark k_1 and k_2, respectively, and we start by considering the real emission of a

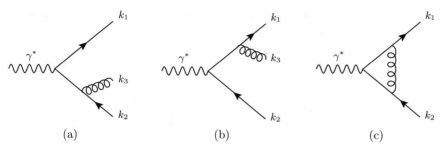

Fig. 2.2 Feynman diagrams contributing to the cross-section of $e^+e^- \to q\bar{q}$ at $\mathcal{O}(\alpha_s)$

soft gluon with momentum k_3, i.e. diagrams in Fig. 2.2a and b. The matrix element for diagram (b) can be written as

$$
\begin{aligned}
M_3^{(b)} &= t_1^a \, g_s \, \bar{u}\,(k_1)\,\gamma^\mu \varepsilon_\mu\,(k_3) \, \frac{\slashed{k}_1 + \slashed{k}_3}{(k_1 + k_3)^2 + i\epsilon} \, \tilde{M}_2 \\
&\xrightarrow{k_3 \to 0} t_1^a \, g_s \, \bar{u}\,(k_1)\,\gamma^\mu \varepsilon_\mu\,(k_3) \, \frac{\slashed{k}_1}{2k_1 \cdot k_3 + i\epsilon} \, \tilde{M}_2 \\
&= t_1^a \, g_s \, \frac{k_1^\mu}{k_1 \cdot k_3} \varepsilon_\mu\,(k_3)\,\bar{u}\,(k_1)\,\tilde{M}_2 = t_1^a \, g_s \, \frac{k_1^\mu}{k_1 \cdot k_3} \varepsilon_\mu\,(k_3)\,M_2, \quad (2.12)
\end{aligned}
$$

where we have used anti-commutation relations of the Dirac matrices and $\slashed{k}_1 u(k_1) = 0$ to get the last line. The factor $k_1^\mu/(k_1 \cdot k_3)$ is called eikonal factor and t_1^a is the colour charge associated to the emission of a gluon off a quark line, i.e. it is a generator of SU(3) in the fundamental representation. We have also used fairly standard notation for the Dirac spinor $\bar{u}(k)$ and for the gluon polarisation vector $\varepsilon_\mu(k_3)$. In the last step the Dirac spinor was absorbed in the 2-parton matrix element M_2 and therefore we dropped the tilde on it. For the full real-emission amplitude we find

$$
M_3 = M_3^{(a)} + M_3^{(b)} \xrightarrow{k_3 \to 0} g_s J^\mu\,(k_3)\,\varepsilon_\mu\,(k_3)\,M_2, \qquad (2.13)
$$

where we have introduced the eikonal current

$$
J^\mu\,(k) = \sum_{i=1}^{2} t_i^a \frac{k_i^\mu}{k \cdot k_i}. \qquad (2.14)
$$

It is important to note that the factorisation does not depend on the internal structure of the amplitude. From the physical point of view, this reflects the fact that the large wavelength of the soft radiation cannot resolve the details of the short distance interactions. However, the proof of this statement to any perturbative orders is highly non-trivial and it heavily relies on gauge invariance.

We now square the amplitude and we arrive at the following factorised expression for the emission of a soft real gluon

$$
|M_3|^2 \xrightarrow{k_3 \to 0} |M_2|^2 g_s^2 J^\mu (k_3) J^\nu (k_3) (-g_{\mu\nu})
$$

$$
= |M_2|^2 g_s^2 \left[-\sum_{i,j} t_i^a t_j^a \frac{k_i \cdot k_j}{(k_i \cdot k_3)(k_j \cdot k_3)} \right]
$$

$$
= |M_2|^2 g_s^2 C_{12} \frac{k_1 \cdot k_2}{(k_1 \cdot k_3)(k_2 \cdot k_3)}, \qquad (2.15)
$$

where we have introduced the effective colour charge

$$
C_{ij} = -2 t_i^a t_j^a. \qquad (2.16)
$$

We note that the effective colour charge is a matrix which has in principle non-zero entries also away from the diagonal. It is easy to show using colour conservation that its structure noticeably simplifies in the case under consideration, because we only have two hard legs which carry colour:

$$
t_1^a + t_2^a = 0 \implies \left(t_1^a\right)^2 + \left(t_2^a\right)^2 = -2t_1^a t_2^a \implies C_{12} = 2C_F, \qquad (2.17)
$$

where all the above equalities are meant to hold when the matrices act on physical states. The effective colour charge turns out to be diagonal also in the case of three hard coloured legs, as we shall see in Sect. 4.3.2, while with four or more hard partons a non-trivial matrix structure emerges. We point out for the interest reader that a general and rather powerful colour-operator formalism to deal with this issue exists [34–38].

The soft approximation can be applied also to the virtual corrections. i.e. to the diagram in Fig. 2.2c. In this limit we can in general neglect powers of the loop momentum k_3 in the numerator. Moreover, in the denominator we can use the fact that $k_3^2 \ll k_i \cdot k_3$. The loop correction to quark-antiquark pair production is therefore proportional to

$$
I = g_s^2 C_F (-i) \int \frac{d^4 k_3}{(2\pi)^d} \frac{\bar{u}(k_1) \gamma^\mu (\slashed{k}_1 + \slashed{k}_3) \gamma^\rho (\slashed{k}_3 - \slashed{k}_2) \gamma_\mu v(k_2)}{[(k_3 + k_1)^2 + i\epsilon][(k_3 - k_2)^2 + i\epsilon][k_3^2 + i\epsilon]}
$$

$$
\to g_s^2 C_F (-i) \int \frac{d^4 k_3}{(2\pi)^d} \frac{(k_1 \cdot k_2) [\bar{u}(k_1) \gamma^\rho v(k_2)]}{[k_3 \cdot k_1 + i\epsilon][-k_3 \cdot k_2 + i\epsilon][k_3^2 + i\epsilon]}, \qquad (2.18)
$$

where we have written the result in $d = 4$ space-time dimensions because we are going to combine it together with the real-emission part, before calculating the divergent integrals.

It is helpful to use the following parametrisation of the four-momenta:

$$k_1^\mu = E_1\,(1,0,0,1)\,, \quad k_2^\mu = E_2\,(1,0,0,-1)\,,$$

$$k_3^\mu = \left(k_3^0, \vec{k}_3\right) \text{ with } \vec{k}_3 = \left(\vec{k}_{3\perp}, k_3^z\right)\,, \tag{2.19}$$

where $\vec{k}_{3\perp}$ is the vectorial transverse loop momentum and $k_{3\perp} \equiv |\vec{k}_{3\perp}|$. We note that

$$k_{3\perp}^2 = \frac{(k_1.k_2)}{2(k_1.k_3)(k_3.k_2)}\,. \tag{2.20}$$

We thus obtain

$$I = g_s^2 C_F(-i) \int \frac{d^3 k_3}{(2\pi)^4} \frac{2\,dk_3^0\,[\bar{u}\,(k_1)\,\gamma^\rho v\,(k_2)]}{\left(k_3^0 - k_3^z + i\epsilon\right)\left(-k_3^0 - k_3^z + i\epsilon\right)\left(k_3^{0\,2} - k_3^{z\,2} - k_{3\perp}^2 + i\epsilon\right)} \tag{2.21}$$

When performing loop-calculations, one usually introduces a regulator, such as for instance dimensional regularisation, and then evaluates the integrals in Eq. (2.21) directly. Here we take another approach which allows us to highlight the similarities between loop integrals for the virtual terms and phase-space integrals for the real contributions. We want first to evaluate the integral in k_3^0. We note that the integrand has four poles in the complex k_3^0 plane, which are located at

$$k_3^0 = k_3^z - i\epsilon, \qquad k_3^0 = -k_3^z + i\epsilon, \qquad k_3^0 = \pm\left(|\vec{k}_3| - i\epsilon\right)\,. \tag{2.22}$$

Closing the contour from below we find

$$I = g_s^2 C_F\left[\bar{u}\,(k_1)\,\gamma^\rho v\,(k_2)\right] \int \frac{d^3 k_3}{(2\pi)^3} \left[\frac{-(k_1 \cdot k_2)}{2|\vec{k}_3|\,(k_1 \cdot k_3)\,(k_2 \cdot k_3)}\right.$$

$$\left. -\frac{1}{\left(k_3^z - i\epsilon\right)\left(k_{3\perp}^2\right)}\right]\,, \tag{2.23}$$

where the second integral is a pure phase

$$\int \frac{dk_3^z\,d^2 k_{3\perp}}{(2\pi)^3} \frac{1}{\left(k_3^z - i\epsilon\right)\left(k_{3\perp}^2\right)} = -\int dk_3^z \frac{k_3^z + i\epsilon}{k_3^{z\,2} + \epsilon^2} \int \frac{dk_{3\perp}}{(2\pi)^2} \frac{1}{k_{3\perp}}$$

$$= -\int \frac{(i\pi)}{(2\pi)^2} \frac{dk_{3\perp}}{k_{3\perp}}\,. \tag{2.24}$$

This contribution is usually referred to as the Coulomb, or Glauber, phase. We note that the above phase always cancels when considering physical cross-sections in Abelian theories like QED. However, it can have a measurable effect in QCD cross-sections, in the presence of a high enough number of harder coloured legs, which lead to a non-trivial matrix structure for the effective colour charges Eq. (2.16).

Collecting real and virtual contributions together, we can compute the NLO distribution of an observable v by introducing an appropriate measurement function $V_n (\{k_i\})$, which describes the value of the observable for a set of n final-state particles $k_1, \ldots, k_n$. The measurement function can contain Dirac delta corresponding to constraints imposed in differential distributions, and/or Heaviside Θ functions, for example when one imposes cuts on the final-state or if one works with cumulative distributions. Furthermore, if we are dealing with jet observables, the measurement functions must also tell us how to combine particles in a jet, i.e. it must specify the jet algorithm (cf. Chap. 3).[5] With this in mind, we can write the cross-section for an observable v to NLO accuracy as the sum of three contribution: Born, real emission and virtual corrections:

$$
\sigma (v) = \frac{1}{2s} \int d\Phi_2 \, |M_2|^2 \, V_2 (k_1, k_2) \tag{2.25}
$$
$$
+ \frac{1}{2s} \int d\Phi_2 \, |M_2|^2 \int \frac{d^3 k_3}{(2\pi)^3 2|\vec{k}_3|} 2g_s^2 C_F \frac{(k_1 \cdot k_2)}{(k_1 \cdot k_3)(k_2 \cdot k_3)}
$$
$$
\times [V_3 (k_1, k_2, k_3) - V_2 (k_1, k_2)].
$$

We note that the Born contribution and the one-loop corrections live in the two-particle phase-space and are characterised by the same measurement function. Instead, the real emission contribution live in a three-body phase-space and, consequently, the measurement function is the three-particle one.

The main result of our discussion so far is Eq. (2.25), which describes the behaviour of a typical NLO cross-section in the limit where the radiated parton (gluon) is soft. However, if we take a closer look we note that

$$
k_i \cdot k_3 = E_i E_3 (1 - \cos\theta_{i3}), \quad i = 1, 2. \tag{2.26}
$$

Thus, the eikonal factor exhibits a singularity not only in the soft limit but also when the parton with momentum k_3 becomes collinear with either k_1 or k_2. It is clear that, while the eikonal approximation is sufficient to correctly capture both the soft-collinear and soft wide-angle, we have to extend our formalism in order to include also the relevant hard-collinear terms. It must be noted that the collinear limit is in many respects easier than the soft limit discussed so far, essentially because the collinear factorisation emerges from a semi-classical picture whereby a parent parton splits into two daughters. An important consequence of this fact is

---

[5]Technically, the jet clustering can usually be written as a series of Θ functions.

that collinear singularities are always accompanied by diagonal colour charge C_{ii}, which is the Casimir of the relevant splitting, i.e. C_F for quark splittings and C_A for gluon splittings.[6] The splitting of a quark into a gluon with momentum fraction z and a quark with momentum fraction $1 - z$ $q \rightarrow qg$ is described at LO by the splitting function

$$P_q(z) = C_F \frac{1 + (1 - z)^2}{z}, \tag{2.27}$$

while the gluon splitting into a pair of gluons or a quark-antiquark pair reads

$$P_g(z) = C_A \left[2 \frac{1 - z}{z} + z(1 - z) + \frac{n_f T_R}{C_A} \left(z^2 + (1 - z)^2 \right) \right]. \tag{2.28}$$

where the first contribution describes the splitting $g \rightarrow gg$, while the second one, proportional to $n_f T_R$ with n_f the number of massless flavours, corresponds to the splitting $g \rightarrow q\bar{q}$. We note that both Eqs. (2.27) and (2.28) exhibit a $z \rightarrow 0$ singularity which is the soft singularity of Eq. (2.25), while the finite-z part of the splitting functions describe the hard-collinear contribution.

2.4 Infra-Red and Collinear Safety

We are now ready to discuss infra-red and collinear safety in a more detailed way. Let us go back to Eq. (2.25), or alternatively we could consider its extension in the collinear limit. In order to achieve a complete cancellation of the IRC singularities, we must consider observables V that satisfy the following properties, which we take as the definition of IRC safety [42]:

collinear safety: $V_{m+1} (\ldots, k_i, k_j, \ldots) \longrightarrow V_m (\ldots, k_i + k_j, \ldots)$ if $k_i \parallel k_j$,
$$\tag{2.29}$$

infrared safety: $V_{m+1} (\ldots, k_i, \ldots) \longrightarrow V_m (\ldots, k_{i-1}, k_{i+1}, \ldots)$ if $k_i \rightarrow 0$.
$$\tag{2.30}$$

---

[6]We warn the reader that although physically motivated, this statement is all but trivial to show! After the first splitting the total colour charge will be shared among the two partons and further radiation can be emitted from either of them. This leads to a colour radiation pattern which is in principle rather complicated. However, soft radiation cannot resolve the details of the interaction which happens at shorter distance and higher momentum scale, a phenomenon called *coherence*. Therefore a soft gluon emitted at an angle θ will only see the total colour charge of the radiation emitted at smaller angles [39–41]. The iteration of this argument essentially leads angular-ordered parton showers and to the resummation of large logarithms in the framework of the coherent branching algorithm [15, 16].

In words, whenever a parton is split into two collinear partons, or whenever an infinitesimally soft parton is added—i.e. in situations where an extra emission makes the real amplitude divergent—the value of the observable must remain unchanged, in order to guarantee a proper cancellation of the divergence against virtual corrections. The above limits have to hold not only for a single particle, but for an ensemble of partons becoming soft and/or collinear. IRC safe properties of jet cross-sections and related variables, such as event shapes and energy correlation functions were first studied in Refs. [43–45].

Let us consider first the case of inclusive observables, i.e. observables that do not constrain additional radiation. We then have $V_m(k_1, \ldots, k_m) = 1$ for all m and the cancellation is complete. Consequently, the total cross-section remains unchanged by the emission of soft particles, as it should. Note that Eq. (2.25) is computed in the soft limit. An exact calculation involves additional corrections, non-divergent in the soft limit, so that the NLO contribution is a finite $\mathcal{O}(\alpha_s)$ correction. Finally, and more interestingly for the topic of this book, let us consider the case of an exclusive (but IRC safe) measurement. Although the singularities cancel, the kinematic dependence of the observable can cause an imbalance between real and virtual contributions, which manifests itself with the appearance of potentially large logarithmic corrections to any orders in perturbation theory. As we have previously mentioned, these logarithmic become large if $v \ll 1$, i.e. if the measurement function constrains real radiation in a small corner of phase-space. These contributions spoil the perturbative expansion in the strong coupling and must be resummed to all orders in order to obtain reliable theoretical predictions for exclusive measurements. A typical observable in jet physics is the jet invariant mass m indeed suffers from these large logarithmic corrections, if we are to consider the boosted regime $p_t \gg m$, where p_t is the jet transverse momentum. We will study the jet mass distribution in great detail in Chap. 4 and discuss how its behaviour is modified by jet substructure algorithms called *groomers*, in Chap. 6.

We note here that there exists a wealth of observables that are of great interest despite them being IRC unsafe. Generally speaking, these observables require the introduction of non-perturbative functions to describe their soft and/or collinear behaviour. For example, lepton-hadron and hadron-hadron cross-sections are written as a momentum-fraction convolution of partonic cross-sections and parton distribution functions. Arbitrary collinear emissions change the value of the momentum fraction that enters the hard scattering, resulting in un-cancelled collinear singularities. Finite cross-sections are then obtained by a renormalisation procedure of the parton densities. Similar situations are also encountered in final-state evolution, if one is interested in measuring a particular type of hadron (see e.g. [46]) or if the measurement only involves charged particles [47, 48]. Furthermore, we mention that recent work [49–51] has introduced the concept of Sudakov safety, which enables to extend the reach of (resummed) perturbation theory beyond the IRC domain. We will come back to this in Chap. 9.

2.5 Hadron Collider Kinematics

Although we have so far considered e^+e^- collisions, which provide an easy framework for QCD studies, the majority of this book will focus on hadron-hadron colliders, with the LHC and possible future hadronic colliders in mind. All the concepts and arguments discussed above remain valid either straightforwardly, or with little adjustments. One of these adjustments is the choice of kinematic variables. This is what we discuss in this section, so as to make our notations clear for the rest of this book.

In the factorised picture described earlier, cf. Eq. (2.4), the hard interaction of a hadron-hadron collisions is really an interaction between two high-energy partons, one from each beam. These two partons carry respectively a fraction x_1 and x_2 of the proton's momentum. Since in general, x_1 and x_2 are different, the centre-of-mass of the hard interaction is longitudinally boosted (along the beam axis) compared to the lab frame. We therefore need to use a set of kinematic variables which is well-behaved with respect to longitudinal boosts. Instead of using energy and polar angles, one usually prefers to use *transverse momentum p_t*, *rapidity y* and *azimuthal angle ϕ*. For a four-vector (E, p_x, p_y, p_z), p_t and ϕ are defined as the modulus and azimuthal angle in the transverse plane (p_x, p_y), i.e. we have

$$p_t = \sqrt{p_x^2 + p_y^2}, \qquad (2.31)$$

and rapidity is defined as

$$y = \frac{1}{2} \log\left(\frac{E + p_z}{E - p_z}\right). \qquad (2.32)$$

In other words, a four-vector of mass m can be represented as

$$p^\mu \equiv (m_t \cosh y, p_t \cos\phi, p_t \sin\phi, m_t \sinh y), \qquad (2.33)$$

with $m_t = \sqrt{p_t^2 + m^2}$ often referred to as the transverse mass. As for the e^+e^- case, a particle of mass m is described with one dimensionful (energy-like) variable, p_t, and two dimensionless variables with a cylindrical geometry: y and ϕ. One can then define a distance (extensively used in this book) between two particles in the (y, ϕ) plane:

$$\Delta R_{12} = \sqrt{\Delta y_{12}^2 + \Delta \phi_{12}^2}. \qquad (2.34)$$

Since we shall integrate over particles produced in the final-state, it is helpful to mention that with the above parametrisation, we have

$$\int \frac{d^4k}{(2\pi)^4}(2\pi)\delta(k^2) = \frac{1}{16\pi^2}\int dk_t^2\, dy \int_0^{2\pi} \frac{d\phi}{2\pi} \qquad (2.35)$$

It is a straightforward exercise in relativistic kinematics to show that for two four-vectors of rapidities y_1 and y_2, the difference $y_1 - y_2$ remains invariant upon a longitudinal boost of the whole system. Additionally, if we come back to the two incoming partons carrying respective fractions x_1 and x_2 of the beam energies, it is easy to show that the centre-of-mass of the collisions has a rapidity $y_{\text{collision}} = \frac{1}{2}\log\left(\frac{x_1}{x_2}\right)$ with respect to the lab frame.

Finally, in an experimental context, one often makes use of the *pseudo-rapidity* η instead of rapidity. The former is directly defined either in terms of the modulus $|\vec{p}|$ of the 3-momentum, or in terms of the polar angle θ between the direction of the particle and the beam:

$$\eta = \frac{1}{2}\log\left(\frac{|\vec{p}| + p_z}{|\vec{p}| - p_z}\right) = -\log\left(\tan\frac{\theta}{2}\right). \qquad (2.36)$$

Contrary to rapidity differences, pseudo-rapidity differences are generally *not* invariant under longitudinal boosts, meaning that one should use rapidity whenever possible. For massless particles $y = \eta$ but this does not hold for massive particles. Hence, for a final-state of massless particles pseudo-rapidity and rapidity can be swapped, but they differ for more complex objects like jets (see next chapter) which have acquired a mass. For these objects, it is recommended to use rapidity whenever possible.

Chapter 3
Jets and Jet Algorithms

3.1 The Concept of Jets

When studying high-energy collisions one often has to consider processes where quarks and gluons are produced in the final-state. For e^+e^- collisions, the study of hadronic final-states has been a major source of information, helping to establish QCD as the fundamental theory of strong interactions, but also providing a clean playground for the study of perturbative QCD and the tuning of Monte-Carlo event generators. At the LHC, the list of processes involving high-energy quarks and/or gluons in their final state is even longer. First, since we collide protons, a hard QCD parton can be radiate from the incoming partons. Then, other particles like W, Z and Higgs bosons can themselves decay to quarks. And, finally, when searching for new particles, one often has to consider decay chains involving quarks and gluons.

However, these high-energy quarks and gluons are not directly observed in the final state of the collision. First of all, as mentioned in the previous chapters, they tend to undergo successive branchings at small angles, producing a series of collimated quarks and gluons. The fact that this parton shower is collimated traces back to the collinear divergence of QCD. Starting from a parton with high virtuality (of the order of the hard scale of the process), the parton shower will produce branchings into further partons of decreasing virtuality, until one reaches a non-perturbative (hadronisation) scale, typically of order Λ_{QCD} or 1 GeV. At this stage, due to confinement, these quarks and gluons will form hadrons. Although some analytic approaches to hadronisation exist, this non-perturbative step often relies on models implemented in Monte Carlo Event generators.

Overall, the high-energy partons produced by the collision appear in the final state as a collimated bunch of hadrons that we call *jets*. Conceptually, *jets are collimated flows of hadrons and they can be seen as proxies to the high-energy quarks and gluons produced in a collision*. This behaviour is observed directly in experiments where the hadronic final state appears to be collimated around a few directions in the detector.

© Springer Nature Switzerland AG 2019
S. Marzani et al., *Looking Inside Jets*, Lecture Notes in Physics 958,
https://doi.org/10.1007/978-3-030-15709-8_3

3.1.1   Jet Definitions and Algorithms

The above picture is over-simplified in a few respects. First of all, partons are ill-defined objects, e.g. due to higher-order QCD corrections where additional partons, real or virtual, have to be included. Then, whether two particles are part of the same jet or belong to two separate jets also has some degree of arbitrariness, related to what we practically mean by "collimated".

The simple concept of what a jet is meant to represent is therefore not sufficient to practically identify the jets in an event. To do that, one relies on a *jet definition*, i.e. a well-defined procedure that tells how to reconstruct the jets from the set of hadrons in the final state of the collision.

A jet definition can be seen as made of a few essential building blocks: the *jet algorithm*, which is the recipe itself and a set of parameters associated with free knobs in the algorithm. A typical parameter, present in almost all jet definitions used in hadron colliders is the *jet radius* which essentially provides a distance in the rapidity-azimuth ($y - \phi$) plane above which two particles are considered as no longer part of the same jet, i.e. no longer considered as collinear.

In addition, a jet definition uses a *recombination scheme* which specifies how the kinematic properties of the jet are obtained from its constituents. Most applications today use the "*E-scheme*" recombination scheme which simply sums the components of the four-vectors. Other recombination schemes, like the massless p_t or E_t schemes, have been used in the past but are not discussed here. Several jet-substructure applications make use of the *winner-take-all* (WTA) recombination scheme [52] where the result of the recombination of two particles has the rapidity, azimuth and mass of the particle with the larger p_t, and a p_t equal to the sum of the two p_t's. As we will further discuss later in this book, this approach has the advantage that it reduces effects related to the recoil of the jet axis when computing jet observables that share similarities with the event-shape broadening [53].

Over the past few decades, a number of jet algorithms have been proposed. They typically fall under two big categories: *cone algorithms* and *sequential-recombination algorithms*. We discuss them both separately below, focusing on the algorithms that have been most commonly used recently at hadronic colliders. For an extensive review on jet definitions, we highly recommend the reading of Ref. [54].

3.1.2   Basic Requirements

Before giving explicit descriptions of how the most commonly-used jet algorithms are defined, we briefly discuss what basic properties we do expect them to satisfy. In the 1990s a group of theorists and Tevatron experimentalists formulated what is known as the Snowmass accord [55]. This document listed the fundamental criteria that any jet algorithm should satisfy.

Several important properties that should be met by a jet definition are:

1. Simple to implement in an experimental analysis;
2. Simple to implement in the theoretical calculation;
3. Defined at any order of perturbation theory;
4. Yields finite cross sections at any order of perturbation theory;
5. Yields a cross section that is relatively insensitive to hadronisation.

The first two criteria are mostly practical aspects. For example, if an algorithm is too slow at reconstructing jets in an experimental context, it would be deemed impractical. These two conditions also mean that the algorithm should be applicable to an input made either of partons (in a theoretical calculation), or of tracks and calorimeter towers (in an experiment analysis). The third and fourth conditions are mainly those of IRC safety, a requirement that, as we have already seen, is at the core of perturbative QCD calculations. The fifth condition is a little bit more subjective. We have already seen that the description of a particle-collision event relies upon several building blocks: the short-distance interaction computed in fixed-order perturbation theory, the parton shower, the hadronisation process and multi-parton interactions. Since jets are supposed to capture the "hard partons in an event", one should hope that the jets which come out of each of these different steps of an event simulation are in good agreement. In particular, this means that observables built from jet quantities should be as little sensitive as possible to non-perturbative effects like hadronisation and the Underlying Event. Furthermore, to be simple to implement in an experimental analysis, the jets should also be as little sensitive as possible to detector effects and pileup.

The question of the sensitivity of different jet definitions to non-perturbative effects, pileup and detector effects has been an active topic of discussion when deciding which algorithm to use at Tevatron and the LHC. A complete assessment of this question is clearly beyond the scope of the present lecture notes. We will however come back to a few crucial points when introducing the different relevant jet definitions below.

3.2 Sequential Recombination Algorithms

Sequential recombination algorithms are based on the concept that, from a perturbative QCD viewpoint, jets are the product of successive parton branchings. These algorithms therefore try to invert this process by successively recombining two particles into one. This recombination is based on a distance measure that is small when the QCD branching process is kinematically enhanced. Thus, one successively recombine particles which minimise the distance in order to mimic the

QCD dynamics of the parton shower. It is easy to check that all the recombination algorithms described below are infrared-and-collinear safe.

Generalised-k_t Algorithm   Most of the recombination algorithms used in the context of hadronic collisions belong to the family of the *generalised-k_t algorithm* [56] which clusters jets as follows.

1. Take the particles in the event as our initial list of objects.
2. From the list of objects, build two sets of distances: an *inter-particle distance*

$$d_{ij} = \min(p_{t,i}^{2p}, p_{t,j}^{2p})\Delta R_{ij}^2, \tag{3.1}$$

where p is a free parameter and ΔR_{ij} is the geometric distance in the rapidity-azimuthal angle plane (Eq. (2.34)), and a *beam distance*

$$d_{iB} = p_{t,i}^{2p} R^2, \tag{3.2}$$

with R a free parameter usually called the *jet radius*.
3. Iteratively find the smallest distance among all the d_{ij} and d_{iB}

- If the smallest distance is a d_{ij} then objects i and j are removed from the list and recombined into a new object k (using the recombination scheme) which is itself added to the list.
- If the smallest is a d_{iB}, object i is called a *jet* and removed from the list.

Go back to step 2 until all the objects in the list have been exhausted.

In all cases, we see that if two objects are close in the rapidity-azimuth plane, as would be the case after a collinear parton splitting, the distance d_{ij} becomes small and the two objects are more likely to recombine. Similarly, when the inter-particle distances are such that $\Delta R_{ij} > R$, the beam distance becomes smaller than the inter-particle distance and objects are no longer recombined, making R a typical measure of the size of the jet.

k_t Algorithm   Historically, the best-known algorithm in the generalised-k_t family is the k_t *algorithm* [57, 58], corresponding to $p = 1$ above. In that case, a soft emission, i.e. one with small p_t, would also be associated a small distance and therefore recombine early in the clustering process. This is motivated by the fact that soft emissions are also enhanced in perturbative QCD.[1] Its sensitivity to soft emissions, while desirable from a perturbative QCD standpoint, has the disadvantage that jets become more sensitive to extra soft radiation in the event, typically like the Underlying Event or pileup. Although the Tevatron experiments have sometimes resorted to the k_t algorithm, they have predominantly used cone algorithms (see below) for that reason.

---

[1]Note that the presence of the "min" in the distance measure, instead of a product, guarantees that two soft objects far apart are not recombined. This would lead to undesired behaviours and complex analytic structures, as it is the case with the JADE algorithm [59, 60].

Cambridge/Aachen Algorithm Another specific cases of the generalised-k_t algorithm is the *Cambridge/Aachen algorithm* [61, 62], obtained by setting $p = 0$ above. In this case, the distance becomes purely geometrical and suffers less from the contamination due to soft backgrounds than the k_t algorithm does.

Anti-k_t Algorithm In the context of LHC physics, jets are almost always reconstructed with the *anti-k_t algorithm* [63], which corresponds to the generalised-k_t algorithm with $p = -1$. The primary advantage of this choice is that it favours hard particles which will cluster first. A hard jet will grow by successively aggregating soft particles around it until it has reached a (geometrical) distance R away from the jet axis. This means that hard jets will be insensitive to soft radiation and have a circular shape in the $y - \phi$ plane. This soft-resilience of the anti-k_t algorithm largely facilitates its calibration in an experimental context and is the main reason why it was adopted as the default jet clustering algorithm by all the LHC experiments.

To make things more concrete, we show in Fig. 3.1 a step-by-step example of a clustering sequence with the anti-k_t jet algorithm on a small set of particles. The successive pairwise recombinations, and beam recombination giving the final jets, is clearly visible on this figure. Finally, the resilience of anti-k_t jets with respect to soft radiation is shown in Fig. 3.2, where we see that anti-k_t jets have a circular shape while Cambridge/Aachen jets have complex boundaries.[2]

Relevance for Jet Substructure In the context of jet substructure studies, several recombination algorithms are used. Initially, jets are usually reconstructed using the anti-k_t algorithm with a large radius (typically R in the 0.8–1.2 range). Many substructure tools then rely on reclustering the constituents of that jet with another sequential-recombination jet algorithm (or jet definition), allowing one to have a convenient view of the jet clustering as a tree structure. The most commonly used algorithm is probably Cambridge/Aachen since it gives a natural handle on the structure of the jet at different angular scales, in a way that respects the angular ordering of parton showers (see also [65]). One also relies on the k_t algorithm used e.g. to split the jet into subjets, or the generalised-k_t algorithm with $p = 1/2$, used because it mimics an mass/virtuality ordering of the subjets. More details will be given later when we review the main substructure tools.

3.3 Cone Algorithms

Cone algorithms were first introduced in 1979 [42]. They are based on the idea that jets represent dominant flows of energy in an event. Modern cone algorithms rely on the concept of a *stable cone*: for a given cone centre y_c, ϕ_c in the rapidity-

---

[2]In practice, the jet areas are obtained by adding a infinitely soft particles, *aka* ghosts, to each calorimeter tower, These are clustered with the hard jets, indicating the boundaries of the jets.

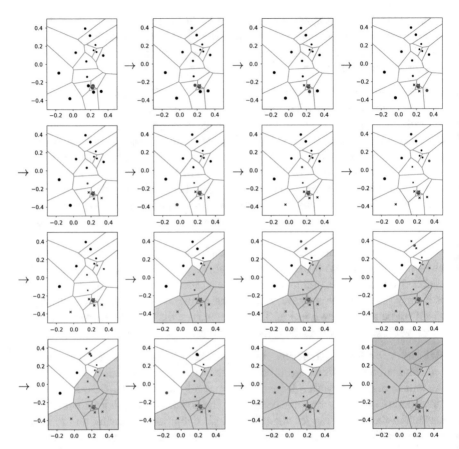

Fig. 3.1 Illustration of a step-by-step clustering using the anti-k_t algorithm with $R = 0.4$. The axes of each plot are rapidity and azimuthal angle. Each particle is represented by a cross with a size increasing with the p_t of the particle. To help viewing the event, we also draw in grey lines the Voronoi cells obtained for the set of particles in the event (i.e. cells obtained from the bisectors of any pair of points). Each panel corresponds to one step of the clustering. At each step, the dots represent the objects which are left for clustering (again, with size increasing with p_t). Pairwise clusterings are indicated by a blue pair of dots, while red dots correspond to final jets (i.e. beam clusterings). The shaded areas show the cells included in each of the three jets which are found ultimately

azimuth plane, one sums the 4-momenta of all the particles with rapidity and ϕ within a (fixed) radius R around the cone centre; if the 4-momentum of the sum has rapidity y_c and azimuth ϕ_c—i.e. the sum of all the momenta in the cone points in the direction of the centre of the cone—the cone is called *stable*. This can be viewed as a self-consistency criterion.

In order to find stable cones, the JetClu [66] and (various) midpoint-type [67, 68] cone algorithms use a procedure that starts with a given set of seeds. Taking each of them as a candidate cone centre, one calculates the cone contents, find a new

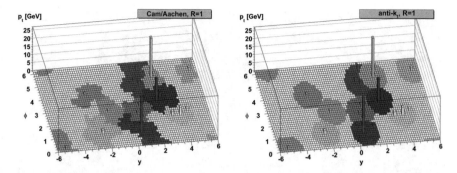

Fig. 3.2 Jets obtained with the Cambridge/Aachen (left) and anti-k_t (right) algorithms with $R = 1$. The shaded regions correspond to the (active) catchment area (see [64]) of each jet. While the jets obtained with the Cambridge/Aachen algorithm have complex boundaries (a similar property would be seen on k_t jets), the hard jets obtained with anti-k_t clustering are almost perfectly circular. This figure has been taken from [63]

centre based on the 4-vector sum of the cone contents and iterate until a stable cone is found. The JetClu algorithm, used during Run I at the Tevatron, takes the set of particles as seeds, optionally above a given p_t cut. This can be shown to lead to an infrared unsafety when two hard particles are within a distance $2R$, rendering JetClu unsatisfactory for theoretical calculations.

Midpoint-type algorithms, used for Run II of the Tevatron, added to the list of seeds the midpoints between any pair of stable cones found by JetClu. This is still infrared unsafe, this time when 3 hard particles are in the same vicinity, i.e. one order later in the perturbative expansion than the JetClu algorithm. This infrared-unsafety issue was solved by the introduction of the SISCone [69] algorithm. It provably finds all possible stable cones in an event, making the stable cone search infrared-and-collinear safe.

Finally, note that finding the stable cones is not equivalent to finding the jets since stable cones can overlap. The most common approach is to run a split–merge procedure once the stable cones have been found. This iteratively takes the most overlapping stable cones and either merges them of splits them depending on their overlapping fraction.

3.4 Experimental Aspects

The experimental input to the jet algorithms previously discussed is reconstructed from energy deposits of elementary particles within the different detector components. The details of the reconstruction differ between the four LHC experiments, e.g. ATLAS uses topoclusters and CMS uses particle-flow objects as inputs to

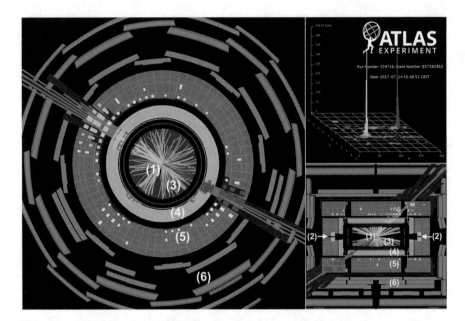

Fig. 3.3 Display of a dijet event recorded by ATLAS in proton-proton collisions at centre-of-mass energy 13 TeV. The two high-p_t jets have both transverse momentum of 2.9 TeV and the dijet system exhibit an invariant mass of 9.3 TeV. The different panels correspond to the view of the event in the plane transverse to the beam direction (large figure on the left-hand side). The two smaller figures on the right-hand side show the calorimeter clusters transverse energies in the (η, ϕ) plane on the top and the longitudinal view of the event on the bottom. The numbers corresponds to different detectors components, as discussed in the text. ATLAS Experiment © 2018 CERN

their jet recombination algorithms.[3] While details of how jet constituents are reconstructed can affect the properties of the jets, we will constrain our discussion here to a generic description of qualitative features in the process of measuring them.

Multi-purpose detectors at the LHC are cylinder-shaped highly-complex objects consisting of layers of different components, as depicted in Fig. 3.3, each component measuring a certain way a particle can interact with the detector. Figure 3.3 shows a dijet event with an invariant mass of the two jets of $m_{jj} = 9.3$ TeV, measured by ATLAS and consists of three different images. In the large image on the left the detector plane transverse to the beam axis is shown. In the lower image on the right we see a lengthwise slice of the ATLAS detector. The upper image on the right shows the energy deposits of particles transverse to the beam axis in the so-called *lego-plot* plane. In the lego plot the cylinder shape of the detector is projected onto a 2-dimensional plane, consisting of the variables $\eta \in (-\infty, \infty)$,

---

[3] ATLAS decided to use particle-flow objects in future studies as well. It will be the default during Run 3 of the LHC.

the pseudo-rapidity, cf. Eq. (2.36), and the azimuthal angle $\phi \in [0, 2\pi]$. η measures how forward a particle is emitted during the proton-proton interaction. Note the similarities between the pseudo-rapidity and the rapidity defined in Eq. (2.32): the two coincide for massless particles. Distances between two cells or particles i and j on the lego plane are measured via

$$\Delta R_{ij}^{(\text{detector})} = \sqrt{(\phi_i - \phi_j)^2 + (\eta_i - \eta_j)^2}. \tag{3.3}$$

Note that the topoclusters are assumed massless, i.e. their rapidity equates their pseudo-rapidity. Thus, for detector cells the definitions of Eqs. (3.3) and (2.34) agree. The different detector components are labelled in Fig. 3.3 in the following way:

(1) Interaction point of the proton beams.
(2) The arrows indicate the direction of the particle beams. The proton beams are entering from either side of the detector and exit on the opposite side after crossing at the collision point.
(3) The innermost part of the ATLAS and CMS detectors consists of the tracking detectors which measure the momentum of charged particles. Strong magnetic fields bend the particles when traversing through the detectors. The way the tracks are bent is indicative of the particle's charge, mass and velocity.
(4) The electromagnetic calorimeter measures predominantly the energies of electrons and photons. Such particles are stopped and induce a cascade of particles, *a shower*, in the calorimeter. Charged particles can be discriminated from photons by the presence or absence of tracks in the tracking detectors. Cell sizes for this calorimeter vary between the central and forward direction of the detector. In the central part they are roughly (0.025×0.025) in the $\phi - \eta$ plane.
(5) The hadronic calorimeter measures the energies of hadronic particles, e.g. protons and neutrons. As in the case of the electromagnetic calorimeter, charged hadrons can be discriminated from neutral ones due to their energy loss in the tracking detectors. The cells that make the hadronic calorimeter have in the central region of the detector a size of roughly (0.1×0.1) in the $\phi - \eta$ plane.
(6) The most outer layer of the detector is the muon spectrometer. Muons, produced with characteristic LHC energies, are weakly interacting with the detector material and are consequently not stopped. However, they may leave tracks in the tracking system, undergo energy loss in the electromagnetic and hadronic calorimeter and may eventually interact with the muon spectrometer.

In Fig. 3.4 we show a segment of a slice of the transverse plane and how classes of particles interact with the individual detector components. For each high-energy Standard Model event we expect of $\mathcal{O}(500)$ resulting particles, which we can classify into photons, charged leptons, neutral and charged hadrons and non-interacting particles, i.e. neutrinos. In a typical proton-proton collision, about 65% of the jet energy is carried by charged particles, 25% by photons, produced mainly

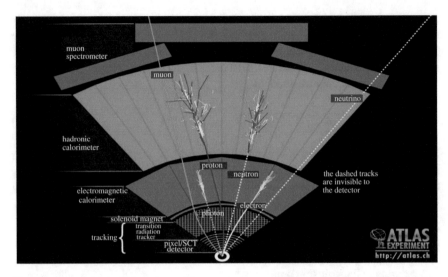

Fig. 3.4 Schematic depiction of a multi-purpose detector, here ATLAS. The picture illustrates how different particles interact with the various layers of the detector. ATLAS Experiment © 2018 CERN

from π_0 decays, and only 10% by neutral hadrons (mostly neutrons and K_L) [70, 71]. However, these fractions can vary significantly from event to event.

Charged particles loose energy when traversing the detector material in various ways. One mechanism is ionisation and excitation interactions with the detector material, e.g. $\mu^- + \text{atom} \to \text{atom}^* + \mu^- \to \text{atom} + \gamma + \mu^-$, where their energy loss per distance is governed by the Bethe equation [72]. Further mechanisms for charged particles to interact with the detector material are *bremsstrahlung*, *direct electron-pair production* and *photonuclear interactions*. Photons interact with the detector material through *photoelectric effect*, *Compton scattering* and *electron-pair production*. The latter being dominant for $E_\gamma \gg 1$ MeV. In the case of hadron-detector interactions, we are dealing mostly with inelastic processes, where secondary strongly interacting particles are produced in the collision.

Information gathered from the detector components (3)–(6) allow to obtain a global picture of the particles produced in the event. However, particles are not directly used as input to construct jets using the algorithms previously discussed. ATLAS and CMS use different approaches to construct jet constituents. The former is using topological clusters, or, in short, topoclusters, which are mainly based on calorimeter objects, while the latter use so-called particle flow objects, which combine information from the tracker and the calorimeter to build a coherent single object.[4] The benefit of using calorimeter objects is a good calibration of the energy component of the topoclusters. On the other hand, the cell size of the hadronic

---

[4]Note that ATLAS is moving to using a particle flow approach as well.

Fig. 3.5 The figure shows how calorimetric information is used by ATLAS to construct jet constituents (taken from [74])

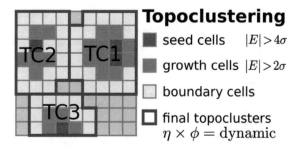

calorimeter is 0.1×0.1 in (η, ϕ) and topological cell clusters are formed around seed cells with an energy $|E_{\text{cell}}|$ at least 4σ above the noise by adding the neighbouring cells with $|E_{\text{cell}}|$ at least 2σ above the noise, and then all surrounding cells [73], see Fig. 3.5. The minimal transverse size for a cluster of hadronic calorimeter cells is therefore 0.3×0.3 and is reached if all significant activity is concentrated in one cell. Two energy depositions leave distinguishable clusters if each one hits only a single cell and their individual axes are separated by at least $\Delta R = 0.2$, so that there is one empty cell between the two seed cells. In the context of this book, it means that if important characteristics of the substructure in a jet are so close that it does not leave separate clusters in the jet, it is impossible to resolve it. This leaves a residual lower granularity scale when using topocluster as fundamental objects to form jets. Thus, in particular when a fine-grained substructure in the jet is of importance, e.g. in the reconstruction of highly boosted resonances, the benefit of particle flow objects is widely appreciated across both multi-purpose experiments.

Focusing exclusively on the tracking detectors when reconstructing jets is an even more radical approach to optimising the spatial resolution of a final state. Tracking detectors can reconstruct the trajectories of a charged particles, which carry $\sim 65\%$ of the final state's energy, and can specify the direction of the particle at any point of the trajectory with a precision much better than the granularity of the calorimeter. For example, the angular resolution of the ATLAS inner tracking detector for charged particles with $p_T = 10$ GeV and $\eta = 0.25$ is $\sim 10^{-3}$ in η and ~ 0.3 mrad in ϕ [75] with a reconstruction efficiency of $> 78\%$ for tracks of charged particles with $p_T > 500$ MeV [76]. Further, the momentum resolution for charged pions is 4% for momenta $|p| < 10$ GeV, rising to 18% at $|p| = 100$ GeV [75]. Note that, generally speaking, the energy resolution tends to degrade with energy in for calorimeters, but improves with energy for trackers.

3.5 Implementation

Most of the practical applications of jets use numerical inputs, either from (fixed-order or parton-shower) Monte Carlo simulations, or directly from experimental data. It is therefore important to have a numerical implementation of the jet

Fig. 3.6 Average clustering time as a function of the event multiplicity N, obtained with the FastJet implementation of several representative algorithms

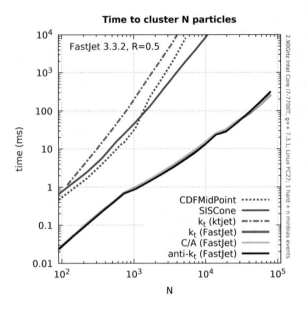

algorithms. Furthermore, this implementation needs to be fast enough for practical usability in an experimental (and, to a lesser extent, theoretical) context. Currently, the standard package for jet clustering is FastJet [56, 77],[5] used by both the experimental and theoretical communities at the LHC. It provides a native implementation of all the recombination algorithms introduced in Sect. 3.2 and plugins for a series of other jet algorithms, including the cone algorithms discussed in Sect. 3.3. As an illustration, we show in Fig. 3.6 the average time it takes to cluster an event with N particles for a few representative algorithms. For the specific case of the k_t algorithm, we show the timings for two different implementations: the initial ktjet implementation [78] available at the time of the Tevatron and deemed too slow, and the FastJet implementation which is faster by 2–3 orders of magnitude in the region relevant for phenomenology (around a few thousands particles). Regarding cone algorithms, this plot shows that infrared-and-collinear SISCone has clustering times similar to the unsafe MidPoint.[6] Finally, if one keeps in mind that in practical (trigger-level) jet reconstruction at the LHC, one has a few tens of milliseconds for clustering, Fig. 3.6 shows that the recombination algorithms (and their FastJet implementation) are currently clearly preferred.

---

[5]See also http://fastjet.fr.

[6]MidPoint has here been used with a seed threshold of 1 GeV. Without a seed threshold, it would be slower by about an order of magnitude.

Chapter 4
Calculations for Jets: The Jet Mass Distribution

In this chapter we begin our discussion about the calculation of jet properties in perturbative QCD. We start by considering an important observable in jet physics, namely the jet invariant mass

$$m^2 = \left(\sum_{i \in \text{jet}} k_i \right)^2, \tag{4.1}$$

where the sum runs over all the particles i which are clustered in the jet. In this lecture notes, because of its simple definition, we are going to take the jet mass as the prototype of a jet substructure observable. This observable will be discussed in detail in this chapter and we will again come back to it in Chap. 6 where we are going to compute the jet mass for jets modified by substructure techniques, a case particularly relevant for phenomenological applications at the LHC.

In our discussion, we shall focus on QCD jets, i.e. jets which are initiated by a hard parton and subsequently evolve through parton shower. Our perturbative analysis will mostly performed at parton level, i.e. we will consider quarks and gluons to be the jet's constituents. Perturbation theory is not able to describe the transition to particle level and hadronisation models are usually employed in event generators to describe the parton-to-hadron transition. In this chapter, we will only briefly comment on these non-perturbative issues, postponing a numerical analysis of their impact to Chap. 6. Even if we remain within the regime of perturbative QCD, we will see that the fixed-order methods are not adequate in order to capture the relevant dynamics of the jet mass, especially in the boosted regime where emissions are accompanied by large logarithms. Thus, we will exploit all-order resummation techniques to better handle the theoretical description of this observable. In order to maintain our presentation as simple as possible, while discussing most of the relevant features, we are going to still focus our discussion on jets produced in e^+e^-. We shall comment on the complication that arise when considering hadron-hadron collisions in Sect. 4.3. In order to make the connection between the e^+e^-

© Springer Nature Switzerland AG 2019
S. Marzani et al., *Looking Inside Jets*, Lecture Notes in Physics 958,
https://doi.org/10.1007/978-3-030-15709-8_4

and the pp discussion as close as possible, we consider in both cases jets clustered with a generalised k_t algorithm with radius R, in its e^+e^- and pp adaptations, respectively [56].

4.1   The One-Loop Calculation

We start by considering the so-called cumulative distribution, which is defined as the normalised cross-section for measuring a value of the jet mass below a certain m^2:

$$\Sigma(m^2) = \frac{1}{\sigma_0} \int_0^{m^2} dm'^2 \frac{d\sigma}{dm'^2} = 1 + \alpha_s \Sigma^{(1)} + \mathcal{O}\left(\alpha_s^2\right), \tag{4.2}$$

where following common practice in the literature, we have chosen to use the Born cross-section as a normalisation factor. The cumulative distribution is a dimensionless quantity and so we can anticipate that its dependence on the jet mass must come as a ratio to another energy scale, which is typically the jet energy (or in proton-proton collision the jet transverse momentum).

We first tackle the calculation of Eq. (4.2) to $\mathcal{O}(\alpha_s)$, in the soft limit. Thus, we consider the eikonal factor for the quark-antiquark dipole (cf. Eq. (2.25))

$$W_{12} = \frac{\alpha_s}{2\pi} (2C_F) \frac{k_1 \cdot k_2}{(k_1 \cdot k_3)(k_2 \cdot k_3)}, \tag{4.3}$$

where k_1 and k_2 are the momenta of the quark and antiquark respectively and k_3 is the momentum of the soft gluon. For instance, we can choose to parametrise them as

$$k_1 = \frac{Q}{2}(1, 0, 0, 1), \quad k_2 = \frac{Q}{2}(1, 0, 0, -1),$$

$$k_3 = \omega(1, \sin\theta\cos\phi, \sin\theta\sin\phi, \cos\theta). \tag{4.4}$$

In terms of the above parametrisation of the kinematics, the Lorentz-invariant phase-space becomes

$$\int d\Phi \equiv \int_0^\infty \omega\, d\omega \int_{-1}^1 d\cos\theta \int_0^{2\pi} \frac{d\phi}{2\pi}. \tag{4.5}$$

This is equivalent to $\frac{1}{\pi} \int d^4k_3 \delta(k_3^2)$, which for simplicity has a slightly different normalisation convention than Eq. (2.35).[1] Note that in the above expression we are

---

[1] Watch out that different conventions are present in the literature. For example (see e.g. [79]), one sometimes uses $\int d^4k_3 \delta(k_3^2)$ as a phase-space integration, in which case Eq. (4.3) has a $\frac{\alpha_s}{2\pi^2}$ factor instead of $\frac{\alpha_s}{2\pi}$.

allowed to ignore any recoil of the quarks against the gluon because we work in
the soft limit. Furthermore, in this limit, the energy of the jet is simply $Q/2$. The
colour factor $2C_F$ in Eq. (4.3) emerges because we are in the presence of only one
dipole. For a process with more hard partonic lines, we should sum over all possible
dipoles each of which is accompanied by the effective colour factors introduced in
Eq. (2.16).

The one-loop evaluation of the cumulative distribution is then obtained adding
together real and virtual corrections. At one loop these contributions are both given
by the eikonal factor W_{12}, but with opposite sign. Another crucial difference is that
when the emitted gluon is real, then we have to impose the appropriate phase-space
constraints. In particular, if the gluon is clustered in the jet seeded by the hard parton
k_1, then its contribution to the jet mass is constrained to be less than m^2. If instead
it falls outside the jet, then it only contributes to the zero-mass bin. In formulae, we
have[2]

$$
\alpha_s \Sigma^{(1)}(m^2) = \int_{-1}^{1} d\cos\theta \int_0^{2\pi} \frac{d\phi}{2\pi} \int_0^Q \omega d\omega \frac{2C_F \alpha_s}{\pi} \frac{1}{\omega^2(1-\cos\theta)(1+\cos\theta)}
$$
$$
\times \left[\Theta_{\text{in jet}} \Theta\left(\frac{Q\omega}{2}(1-\cos\theta) < m^2 \right) + \Theta_{\text{out jet}} - 1 \right]
$$
$$
= -\frac{2\alpha_s C_F}{\pi} \int_{\cos R}^{1-2\frac{m^2}{Q^2}} \frac{d\cos\theta}{(1-\cos\theta)(1+\cos\theta)} \log\left(\frac{Q^2(1-\cos\theta)}{2m^2} \right),
$$
$$
\tag{4.6}
$$

In the above equation, we have used $\Theta_{\text{in jet}} = \Theta(1-\cos\theta < 1-\cos R)$ and
$\Theta_{\text{out jet}} = 1 - \Theta_{\text{in jet}}$, which, for a jet made up of two particles, is the condition
to be satisfied for any clustering algorithm of the generalised k_t family. We will see
in Sect. 4.2.3 that beyond one loop the details of the clustering algorithm affect the
single-logarithmic structure of the jet mass distribution.

The integral over the gluon angle is fairly straightforward. Since we are interested
in the logarithmic region, we neglect powers of the jet mass divided by the hard
scale Q:

$$
\alpha_s \Sigma^{(1)}(m^2) = -\frac{\alpha_s C_F}{2\pi} \left[\log^2\left(\frac{Q^2}{m^2} \tan^2 \frac{R}{2} \right) - \log^2\left(\cos^2 \frac{R}{2} \right) - 2\text{Li}_2\left(\sin^2 \frac{R}{2} \right) \right]
$$
$$
+ \mathcal{O}\left(\frac{m^2}{Q^2} \right),
$$
$$
\tag{4.7}
$$

---

[2]For simplicity, we introduce the following notation for the Heaviside step function: $\Theta(a > b) \equiv$
$\Theta(a - b)$, $\Theta(a < b) \equiv \Theta(b - a)$, and $\Theta(a < b < c) \equiv \Theta(b - a)\Theta(c - b)$.

which is valid for $\frac{m^2}{Q^2} < \sin^2 \frac{R}{2}$. Thus, we see that the jet mass distribution exhibits a double logarithmic behaviour in the ratio of the jet mass to the hard scale. We note that these logarithmic contributions are large if the characteristic energy scale of the jet is much bigger than the jet invariant mass. This situation is precisely what defines boosted topologies and therefore reaching a quantitative understanding boosted-object phenomenology requires dealing with these potentially large logarithmic corrections. As we discussed before, these double logarithms arise from the emission gluons which are both soft and collinear and we therefore expect their presence to any order in perturbation theory. This $\alpha_s^n L^{2n}$ behaviour jeopardises our faith in the perturbative expansion because the suppression in the strong coupling is compensated by the presence of the potentially large logarithm L. In the next section, we will discuss how to resum this contributions, i.e. how to reorganise the perturbative expansions in such a way that logarithmic contributions are accounted for to all orders. We also note that this is necessary only if we are interested in the region $m^2/Q^2 \ll 1$, where the logarithms are large. In the large-mass tail of the distribution instead $m^2/Q^2 \sim 1$ and fixed-order perturbation theory is the appropriate way to capture the relevant physics. Ideally, we would then match resummation to fixed-order to obtain a reliable prediction across the whole range, as shown for instance in Eq. (2.10).

Before moving to the resummed calculation, we want point out two more considerations. First, we can consider a further simplification to Eq. (4.7), namely we can expand it in powers of the jet radius R, which is appropriate for narrow jets

$$\alpha_s \Sigma^{(1)}(m^2) = -\frac{\alpha_s C_F}{2\pi} \log^2 \left(\frac{Q^2 R^2}{4m^2} \right) + \mathcal{O}\left(R^2 \right) = -\frac{\alpha_s C_F}{2\pi} \log^2 \left(\frac{1}{\rho} \right), \quad (4.8)$$

where we have introduced $\rho = \frac{4m^2}{Q^2 R^2}$. Second, we want to discuss further the collinear limit. The starting point of our discussion so far has been the eikonal factor W_{12} in Eq. (4.3), which means that we have only considered the emission of a soft gluon. However, as we discussed in Sect. 2.3, there is another region of the emission phase-space which can produce logarithmic contributions, namely collinear emissions with finite energy ω. We expect this region to be single-logarithmic with the logarithms originating because of the $\cos\theta \to 1$ singularity of the matrix element. The residue of this singularity is given by the appropriate splitting function $P_i(z)$, with $z = \frac{2\omega}{Q}$, which were given in Eqs. (2.27) and (2.28). Our one-loop result is modified accordingly and we get

$$\alpha_s \Sigma^{(1)}(\rho) = -\frac{\alpha_s C_F}{\pi} \left[\frac{1}{2} \log^2 \left(\frac{1}{\rho} \right) + B_q \log \left(\frac{1}{\rho} \right) \right], \quad (4.9)$$

with

$$B_q = \int_0^1 dz \left[\frac{P_q(z)}{2C_F} - \frac{1}{z} \right] = -\frac{3}{4}. \quad (4.10)$$

The collinear limit is of particular relevance when discussing boosted-objects, as radiation is typically collimated along the jet axis. Furthermore, it is often easier from a computational viewpoint to work in such limit because collinear emissions essentially factorise at the cross-section level, while we need to take into account colour correlation at the amplitude level to correctly describe soft emissions at wide angle. Therefore, unless explicitly stated, from now on, we are going to present first calculations in the collinear (and optionally soft) limit and then comment to their extension to include wide-angle soft emission. However, we stress that in general both contributions are necessary to achieve a given (logarithmic) accuracy in the theoretical description of a processes.

4.2   Going to All Orders

In order to obtain theoretical predictions that can be applied in the regime $\rho \ll 1$, we have to move away from fixed-order predictions and resum parton emission to all orders in perturbation theory. Inevitably, we are only going to scratch the surface of the all-order formalism behind resummed calculations and we encourage the interested readers to study more specialised reviews and the original literature on the topic.

For our discussion, we are going to consider a quark-initiated jet in the presence of many collinear (hard or soft) partons. As discussed above, the complete resummed calculation must also consider soft gluons at large angle, while the soft quarks at large angle do not give rise to logarithmic contributions. Let us begin with some consideration on the observable. We want recast the definition Eq. (4.1) in a form which is suitable for the all-order treatment. In the collinear limit, the angular separation between any two jet constituents is small, so we have

$$m^2 = 2 \sum_{(i<j)\in\text{jet}} k_i \cdot k_j = \sum_{(i<j)\in\text{jet}} \omega_i \omega_j \theta_{ij}^2 + \mathcal{O}\left(\theta_{ij}^4\right). \tag{4.11}$$

Any pair-wise distance can be written in terms of each particle's distance from the jet axis and the azimuth in the plane transverse to the jet axis: $\theta_{ij}^2 = \theta_i^2 + \theta_j^2 - 2\theta_i\theta_j \cos\phi_{ij}$. Substituting the above expression in Eq. (4.11), we obtain

$$m^2 = \frac{1}{2} \sum_{(i,j)\in\text{jet}} \omega_i \omega_j \theta_{ij}^2 = \frac{1}{2} \sum_{(i,j)\in\text{jet}} \omega_i \omega_j \left(\theta_i^2 + \theta_j^2 - 2\theta_i\theta_j \cos\phi_{ij}\right)$$

$$= \sum_{i\in\text{jet}} E_J \omega_i \theta_i^2, \tag{4.12}$$

where $E_J = \sum_{i \in \text{jet}} \omega_i = \frac{Q}{2}$ is the jet energy and we have exploited that for each i,

$$\sum_{j \in \text{jet}} \omega_j \theta_j \cos \phi_{ij} = 0, \tag{4.13}$$

because of momentum conservation along i in the plane transverse to the jet.

As before, we are going to consider the cumulative distribution, i.e. the probability for a jet to have an invariant jet mass (squared) less than m^2. We have to consider three cases. Real emissions that are clustered into the jet do contribute to the jet mass distribution, while real emissions outside the jet, as well as virtual corrections, do not change the jet mass. Thus, the cumulative distribution in this approximation reads:

$$\Sigma(\rho) = \sum_{n=0}^{\infty} \frac{1}{n!} \prod_{i=1}^{n} \int \frac{d\theta_i^2}{\theta_i^2} \int dz_i \, P_q(z_i) \frac{\alpha_s(z_i \theta_i \frac{Q}{2})}{2\pi} \Theta_{i \in \text{jet}} \Theta \left(\sum_{i=1}^{n} z_i \frac{\theta_i^2}{R^2} < \rho \right)$$

$$+ \sum_{n=0}^{\infty} \frac{1}{n!} \prod_{i=1}^{n} \int \frac{d\theta_i^2}{\theta_i^2} \int dz_i \, P_q(z_i) \frac{\alpha_s(z_i \theta_i \frac{Q}{2})}{2\pi} \left[\Theta_{i \notin \text{jet}} - 1 \right], \tag{4.14}$$

where the running coupling is evaluated at a scale which represents the transverse momentum of emission i with respect to the $q\bar{q}$ dipole, in the dipole rest frame, cf. Eq. (2.20). The above expression deserves some comments. In order to derive it, we have exploited the factorisation properties of QCD matrix elements squared in the collinear limit. We note that the $1/n!$ prefactor can be viewed as consequence of (angular) ordering. Furthermore, we note that the argument of the each splitting function is energy fraction z_i. This is true if the fractional energy coming out of each splitting is computed with respect to the parent parton. On the other hand, the energy fraction that enters the observable definition is calculated with respect to the jet energy, which in our approximation coincides with the energy of the initial hard quark $E_J = \frac{Q}{2}$. In the collinear limit, these two fractions are related by a rescaling factor x_i that takes into account the energy carried away by previous emissions $x_i = \prod_{k=1}^{i-1}(1 - z_k)$. However, this rescaling only gives rise to subleading (NNLL) corrections and can therefore be dropped in Eq. (4.14). Furthermore, we have also written the jet clustering condition in a factorised form, essentially assuming $\Theta_{i \in \text{jet}} = \Theta(\theta_i < R)$. If the jet is made up of only two particles, this condition is exact for any member of the generalised k_t clustering family. However, there is no guarantee that such condition can be written in a factorised form, in presence of an arbitrary number of particles. Crucially, the widely used anti-k_t algorithm does exhibit this property in the soft limit. In other words, anti-k_t behaves as a perfectly rigid cone in the soft-limit, where all soft particles are clustered first to the hard core, leading to a factorised expression. This is not true with other jet algorithms, such as the Cambridge/Aachen algorithm and the k_t algorithm, for which corrections to the factorised expression occur at NLL accuracy for soft gluon emissions. We will return to this point in Sect. 4.2.3.

With the above clarifications in mind, we can go back to Eq. (4.14). While the second line of (4.14) is already in a fully factorised form, the Θ-function constraining the observable in the first line spoils factorisation. The way around this obstacle is to consider an appropriate integral representation of the Θ function in order to obtain a factorised expression in conjugate space [16, 80]. In other words, we could compute Mellin moments of the cumulative distribution in order to obtain a factorised expression.

At LL accuracy, where each emission comes with a maximal number of logarithms, one can further assume *strong ordering*, i.e. that the $z_i\theta_i^2$ themselves are strongly ordered. In this case, a single emission strongly dominates the sum and we can write

$$\Theta\left(\sum_{i=1}^{n}\rho_i < \rho\right) \approx \Theta\left(\max_i \rho_i < \rho\right) = \prod_{i=1}^{n}\Theta\left(z_i\rho_i < \rho\right), \qquad \rho_i = z_i\frac{\theta_i^2}{R^2},$$

$$(4.15)$$

The fact that, at LL accuracy, a single emission strongly dominates the jet mass is an important result that we will use extensively through this book.

With the above assumptions, it is now straightforward to perform the sum over the number of emissions

$$\Sigma^{(LL)}(\rho) = -\sum_{n=0}^{\infty}\frac{1}{n!}\prod_{i=1}^{n}\int\frac{d\rho_i}{\rho_i}\int dz_i\, P_q(z_i)\frac{\alpha_s(\sqrt{z_i\rho_i}\frac{QR}{2})}{2\pi}$$

$$\times\left[\Theta(\theta < R)\Theta\left(\rho_i > \rho\right)\right] \qquad (4.16)$$

$$= \exp\left[-\int_{\rho}^{1}\frac{d\rho'}{\rho'}\int dz_i\, P_q(z_i)\frac{\alpha_s(\sqrt{z\rho'}\frac{QR}{2})}{2\pi}\Theta(\theta < R)\Theta\left(\rho_i > \rho\right)\right]$$

$$\equiv \exp\left[-R(\rho)\right].$$

This is an interesting and important result: *the cumulative distribution can be written, at LL accuracy, in an exponential form.* At this accuracy, the exponent is determined by the one-gluon contribution and, in particular, can be interpreted as the virtual one-loop contribution, because of the negative sign, evaluated on the region of phase-space where the real emission is vetoed. The function $R(\rho)$ is usually referred to as the *Sudakov exponent* [81] (or the radiator) and it represent the no-emission probability.[3] From the cumulative distribution, we can immediately obtain

---

[3]Please note that throughout this book, R can either denote the jet radius or the radiator/Sudakov exponent. In context, it should be trivial to tell one from the other.

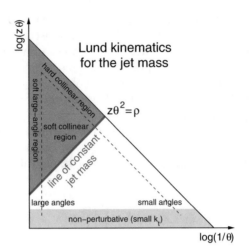

Fig. 4.1 Lund diagram for the jet mass distribution at LL. The solid red line corresponds to emissions yielding the requested jet mass, i.e. with $z\theta^2 = \rho$ (using angles rescaled by R). The shaded red area is the vetoed area associated with the Sudakov suppression. "Soft, wide-angle" emissions have a small k_t and angles of order R, and "hard collinear" splittings have a small angle and a large z fraction. The shaded grey region at the bottom of the plot corresponds to the non-perturbative, small-k_t, region

the resummed jet mass spectrum

$$\frac{\rho}{\sigma_0}\frac{d\sigma}{d\rho} = \frac{d}{d\log(\rho)}\Sigma(\rho) = R'(\rho)e^{-R(\rho)}, \tag{4.17}$$

where $R' = \frac{d}{dL}R$ and $L = \log\left(\frac{1}{\rho}\right)$. It is useful to re-interpret this result in terms of Lund diagrams [82]. These diagrams represent the emission kinematics in terms of two variables: vertically, the logarithm of an emission's transverse momentum k_t with respect to the jet axis, and horizontally, the logarithm of the inverse of the emission's angle θ with respect to the jet axis, (alternatively, we could use its rapidity with respect to the jet axis, if we want to work with hadron colliders coordinates).[4] Note that, in Lund diagrams (and often in actual calculations) we make use of *rescaled* variables, i.e. angles are given in units of the jet radius and the emission transverse momentum (or energy) in units of the jet transverse momentum (or energy). The diagram in Fig. 4.1 shows a line of constant jet mass, together with a shaded (red) region corresponding to the part of the kinematic plane where real emissions are vetoed because they would lead to a value of the mass larger than ρ. In this region, only virtual contributions are allowed, giving rise to the Sudakov

---

[4]More generally, if one considers a gluon emitted from a dipole, as we did in Chap. 2 and earlier in this chapter, one would consider the rapidity along the dipole direction, $-\log(\tan(\theta/2))$, and the transverse momentum $k_\perp$ with respect to the dipole, cf. Eq. (2.20).

factor $\exp[-R(\rho)]$. Outside the shaded (red) region, real and virtual contributions cancel. Because QCD matrix elements are logarithmic in the soft/collinear region, the no-emission probability is proportional to the area of the shaded region (up to running-coupling corrections).

In order to obtain explicit resummed expressions, we have to evaluate the integrals in Eq. (4.16) to the required accuracy. For instance, if we aim to NLL (in the small R limit), we have to consider the running of the strong coupling at two loops. Furthermore, we have to include the complete one-loop splitting function $P_q(z)$ as well its soft contribution at two loops, which corresponds to the two-loop cusp anomalous dimension $K = C_A \left(\frac{67}{18} - \frac{\pi^2}{6} \right) - \frac{5}{9} n_f$. We note that this contribution accounts for correlated gluon emission which are unresolved at NLL accuracy. This correction can therefore be absorbed into the running coupling, giving rise to the so-called Catani-Marchesini-Webber (CMW) scheme [15]:

$$\frac{\alpha_s^{\text{CMW}}(\mu)}{2\pi} = \frac{\alpha_s(\mu)}{2\pi} + K \left(\frac{\alpha_s(\mu)}{2\pi} \right)^2. \tag{4.18}$$

We write the resummed exponent as

$$R(\rho) = L f_1(\lambda) + f_2(\lambda), \tag{4.19}$$

where f_1 and f_2 resum leading and next-to-leading logarithms, respectively:

$$f_1(\lambda) = \frac{C_F}{\pi \beta_0 \lambda} \left[(1 - \lambda) \log (1 - \lambda) - 2 \left(1 - \frac{\lambda}{2} \right) \log \left(1 - \frac{\lambda}{2} \right) \right], \tag{4.20}$$

and

$$\begin{aligned}
f_2(\lambda) = {} & \frac{C_F K}{4\pi^2 \beta_0^2} \left[2 \log \left(1 - \frac{\lambda}{2} \right) - \log (1 - \lambda) \right] - \frac{C_F B_q}{\pi \beta_0} \log \left(1 - \frac{\lambda}{2} \right) \\
& + \frac{C_F \beta_1}{2\pi \beta_0^3} \left[\log (1 - \lambda) - 2 \log \left(1 - \frac{\lambda}{2} \right) \right. \\
& \left. + \frac{1}{2} \log^2 (1 - \lambda) - \log^2 \left(1 - \frac{\lambda}{2} \right) \right],
\end{aligned} \tag{4.21}$$

$\lambda = 2\alpha_s \beta_0 L$, B_q was defined in Eq. (4.10), and $\alpha_s \equiv \alpha_s (QR/2)$ is the $\overline{\text{MS}}$ strong coupling. Since this kind of results will appear repeatedly throughout this book, we give an explicit derivation of the above formulæ in Appendix A. In the above results we have also introduced the one-loop and two-loop coefficients of the QCD β-function, namely β_0 and β_1. Their explicit expressions are given in Appendix A.

In order to achieve the complete NLL resummation formula for the invariant mass distribution of narrow, i.e. small R, jets we need to consider two additional contributions: multiple emissions and non-global logarithms [83]. We have already

mentioned how to deal with the former: in the real-emission contribution to
Eq. (4.14), we can no longer apply the strong-ordering simplification Eq. (4.15)
and the resummed calculation must be done in a conjugate (Mellin) space in order
to factorise the observable definition. At the end of the calculation, the result must
then brought back to physical space. In case of jet masses this inversion can be done
in closed-form and, to NLL accuracy, it can be expressed as a correction factor:

$$\mathcal{M}(\rho) = \frac{e^{-\gamma_E R'(\rho)}}{\Gamma(1 + R'(\rho))}. \tag{4.22}$$

Non-global logarithms are instead resummed into a factor $\mathcal{S}(\rho)$ which has a much
richer (and complex) structure. We will discuss it in some detail in Sect. 4.2.2.
Putting all things together the NLL result for the cumulative mass distribution reads

$$\Sigma^{(\text{NLL})}(\rho) = \mathcal{M}\,\mathcal{S}\,e^{-R}. \tag{4.23}$$

Thus far we have discussed the jet mass distribution in the context of perturbation
theory. However, when dealing with soft and collinear emissions, we are probing the
strong coupling deeper and deeper in the infra-red and we may become sensitive to
non-perturbative contributions. This is clearly dangerous because as the coupling
grows, perturbation theory becomes first unreliable and then meaningless. The
presence of an infra-red singularity (Landau pole) for the coupling makes this
breakdown manifest: at long distances we cannot use partons as degrees of freedom
but we have to employ hadrons. From this point of view it is then crucial to
work with IRC safe observables, for which we can identify regions in which the
dependence on non-perturbative physics can be treated as a (small) correction.

4.2.1 A Sanity Check: Explicit Calculation of the Second Order

As a sanity check of the all-order calculation we have performed in the previous
section, we explicitly calculate the double logarithmic contribution at two loops and
compare it to the expansion of the resummation to second order. Thus, we need
to consider the squares matrix element for the emission of two soft gluons with
momenta k_3 and k_4, off a $q\bar{q}$ dipole, in the limit where both k_3 and k_4 are soft,
with k_4 much softer than k_3 [84, 85]. This can be written as the sum of two pieces:
independent and correlated emissions

$$W = C_F^2 W^{(\text{ind})} + C_F C_A W^{(\text{corr})}, \tag{4.24}$$

where

$$W^{(\text{ind})} = \frac{2\,k_1 \cdot k_2}{k_1 \cdot k_3\, k_2 \cdot k_3} \frac{2\,k_1 \cdot k_2}{k_1 \cdot k_4\, k_2 \cdot k_4}, \tag{4.25}$$

$$W^{(\text{corr})} = \frac{2\,k_1 \cdot k_2}{k_1 \cdot k_3\, k_2 \cdot k_3} \left(\frac{k_1 \cdot k_3}{k_1 \cdot k_4\, k_3 \cdot k_4} + \frac{k_2 \cdot k_3}{k_2 \cdot k_4\, k_3 \cdot k_4} - \frac{k_1 \cdot k_2}{k_1 \cdot k_4\, k_2 \cdot k_4} \right). \tag{4.26}$$

The two contributions are schematically shown in Fig. 4.2. Because we are interested in the $\alpha_s^2 L^4$ contribution to the cumulative distribution, which is the most singular one, we expect it to originate from the independent emission of two gluons in the soft and collinear limit. We have to consider three types of configuration: double real emission, double virtual and real emission at one loop. For each of the three types, the contribution to the squared matrix element for ordered two-gluon emission is the same up to an overall sign. Focusing on the independent emission contribution, the result for the double real (RR) or double virtual (VV) is

$$W^{(\text{ind})} = \frac{256}{Q^4} \frac{1}{z_3^2 z_4^2} \frac{1}{\theta_3^2 \theta_4^2}. \tag{4.27}$$

A similar result holds for the real emission at one loop, with a relative minus sign. The latter has to be counted twice because the real emission could be either k_3 (RV) or the softer gluon k_4 (VR). We are now in a position to compute the jet mass distribution at the two-gluon level for the independent emission C_F^2 term. To perform this calculation we note that it is actually more convenient to consider the differential jet mass distribution rather than the cumulative, as we usually do. In fact, if we demand $m^2 > 0$, then the double virtual configuration does not contribute because it lives at $m^2 = 0$. Therefore we consider

$$\frac{d\tilde{\sigma}}{d\rho} = \frac{1}{\sigma_0} \frac{d\sigma}{d\rho} = \alpha_s \frac{d\tilde{\sigma}^{(1)}}{d\rho} + \alpha_s^2 \frac{d\tilde{\sigma}^{(2)}}{d\rho} + \mathcal{O}\left(\alpha_s^3\right). \tag{4.28}$$

We start by noting that the phase space integration region for all configurations can be divided according to whether the real gluons k_3 and k_4 are inside or outside

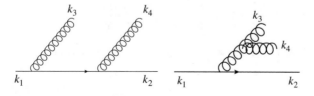

Fig. 4.2 A schematic representation of the types of contributions to the strongly order emission of two soft gluons, in the double real-emission case: independent emission of the left and correlated emission on the right

the jet of interest. We have four distinct regions: k_3, k_4 both outside the jet, k_3, k_4 both inside the jet or either of the gluons inside and the other outside the jet. The condition for a given gluon to end up inside or outside the jet depends on the jet definition. In the anti-k_t algorithm with radius R the condition is particularly simple when considering only soft emissions: a soft emission k_i is inside the jet if it is within an angle R of the hard parton initiating the jet, otherwise it is outside. As we have already noted, the anti-k_t algorithm in the soft limit works as a perfect cone.

Let us consider all four cases one by one. The contribution where both k_3 and k_4 are outside the jet trivially vanishes since it gives a massless jet. We then consider the case where the harder emission k_3 is in the jet and k_4 is out. Graphs RR and RV cancel since the real k_4 does not contribute to the jet mass exactly like the virtual k_4. This leaves diagram VR, which gives zero since the in-jet gluon k_3 is virtual and hence does not generate a jet mass. Hence the region with k_3 in and k_4 out gives no contribution. The contribution where k_4 is in the jet and k_3 out vanishes for the same reason. Hence we only need to treat the region with both gluons in the jet and we shall show that this calculation correctly reproduces the result based on exponentiation of the single gluon result. The sum of the RR, RV and VR contributions can be represented as[5] (with $d\Phi$ defined in Eq. (4.5))

$$\alpha_s^2 \frac{d\tilde{\sigma}^{(2)}}{d\rho} = \int d\Phi \, W \left[\delta \left(\rho - z_3\theta_3^2 - z_4\theta_4^2 \right) - \delta \left(\rho - z_3\theta_3^2 \right) - \delta \left(\rho - z_4\theta_4^2 \right) \right],$$
(4.29)

where in order to keep our notation simple, we have switched to rescaled angular variables: $\theta_i \to \frac{\theta_i}{R}$, so that now $\theta_i < 1$. To proceed, we note that in the leading-logarithmic approximation emissions are also strongly ordered in $z\theta^2$, i.e. we have either $z_3\theta_3^3 \gg z_4\theta_4^2$, or $z_4\theta_4^3 \gg z_3\theta_3^2$. This means that only the largest of $z_3\theta_3^3$ and $z_4\theta_4^2$ contributes to $\delta \left(\rho - z_3\theta_3^2 - z_4\theta_4^2 \right)$, with the other being much smaller. We can therefore write

$$\delta \left(\rho - z_3\theta_3^2 - z_4\theta_4^2 \right) \to \delta \left(\rho - z_3\theta_3^2 \right) \Theta \left(\rho > z_4\theta_4^2 \right) + 3 \leftrightarrow 4 \,.$$
(4.30)

Doing so and using the explicit forms of W and the phase space $d\Phi$ in the small angle limit we get

$$\alpha_s^2 \frac{d\tilde{\sigma}^{(2)}}{d\rho} = - \left(\frac{\alpha_s C_F}{\pi} \right)^2 \int \frac{d\theta_3^2}{\theta_3^2} \frac{d\theta_4^2}{\theta_4^2} \frac{d\phi}{2\pi} \frac{dz_3}{z_3} \frac{dz_4}{z_4}$$
$$\times \left[\delta \left(\rho - z_3\theta_3^2 \right) \Theta \left(z_4\theta_4^2 > \rho \right) + 3 \leftrightarrow 4 \right] \Theta \left(z_3 > z_4 \right),$$
(4.31)

---

[5]Here with an abuse of notation we are indicating the LHS of the equation as $\alpha_s^2 \frac{d\tilde{\sigma}^{(2)}}{d\rho}$, while we really mean only its double leading contribution.

where ϕ is the azimuthal angle between the two gluons (the other azimuthal integration is trivial because the matrix element does not depend on either ϕ_3 or ϕ_4). We note that the overall factor $-\Theta\left(z_4\theta_4^2 > \rho\right)$ comes again from the region where k_4 is virtual, while real and virtual emissions cancel each other for $z_4\theta_4^2 < \rho$. Carrying out the integrals we obtain

$$\alpha_s^2 \frac{d\tilde{\sigma}^{(2)}}{d\rho} = -\frac{1}{2}\left(\frac{\alpha_s C_F}{\pi}\right)^2 \frac{1}{\rho} \log^3\left(\frac{1}{\rho}\right), \tag{4.32}$$

which is precisely the result obtained by expanding the exponentiated double-logarithmic one-gluon result to order α_s^2 and differentiating with respect to ρ. Thus the standard double-logarithmic result for the jet-mass distribution arises entirely from the region with both gluons in the jet. Contributions from soft emission arising from the other regions cancel in the sense that they produce no relevant logarithms.

We note that since we have used a soft-gluon approximation (with gluons emitted from colour dipoles), the result above does not include the contribution from hard-collinear splittings which, at this order would give a contribution $\frac{3}{2}\left(\frac{\alpha_s C_F}{\pi}\right)^2 \frac{1}{\rho} \log^2(\rho) B_q$. Finally, beyond the double-logarithmic approximation, the approximation (4.30) is no longer valid. It does bring a correction to Eq. (4.32) coming from the difference between the left-hand side and the right-hand side of (4.30). In practice, we get

$$\frac{1}{2}\left(\frac{\alpha_s C_F}{\pi}\right)^2 \int \frac{d\theta_3^2}{\theta_3^2}\frac{d\theta_4^2}{\theta_4^2}\frac{dz_3}{z_3}\frac{dz_4}{z_4}\left[\delta\left(\rho - z_3\theta_3^2 - z_4\theta_4^2\right)\right.$$
$$\left. - \delta\left(\rho - z_3\theta_3^2\right)\Theta\left(z_4\theta_4^2 > \rho\right) - 3 \leftrightarrow 4\right]$$
$$= \left(\frac{\alpha_s C_F}{\pi}\right)^2 \int_0^\rho \frac{d\rho_3}{\rho_3}\frac{d\rho_4}{\rho_4}\log\left(\frac{1}{\rho_3}\right)\log\left(\frac{1}{\rho_4}\right)$$
$$\times\left[\delta\left(\rho - \rho_3 - \rho_4\right) - \delta\left(\rho - \rho_3\right)\right]\Theta(\rho_3 > \rho_4)$$
$$= \left(\frac{\alpha_s C_F}{\pi}\right)^2 \frac{1}{\rho}\int_0^\rho \frac{d\rho_4}{\rho_4}\log\left(\frac{1}{\rho_4}\right)\left[\log\left(\frac{1}{\rho - \rho_4}\right) - \log\left(\frac{1}{\rho}\right)\right]$$
$$= \left(\frac{\alpha_s C_F}{\pi}\right)^2 \frac{1}{\rho}\frac{\pi^2}{6}\log\left(\frac{1}{\rho}\right) + (\text{terms with no } \log(\rho) \text{ enhancements})$$

$$\tag{4.33}$$

where we have introduced $\rho_i = z_i\theta_i^2$ and used $\int \frac{d\theta_i^2}{\theta_i^2}\frac{dz_i}{z_i}f(\rho_i) = \int \frac{d\rho_i}{\rho_i}\log(1/\rho_i)$ $f(\rho_i)$. It is easy to show that this contribution corresponds exactly to the first non-trivial correction from $\mathcal{M}(\rho)$ in Eq. (4.22), after differentiation with respect to ρ, with $R'(\rho) = \frac{\alpha_s C_F}{\pi}\log\left(\frac{1}{\rho}\right)$.

4.2.2 Non-global Logarithms

In Sect. 4.2 we have described an all-order calculation that aims to resum large logarithms of the ratio of the jet mass to the hard scale of the process to NLL. Furthermore, in Sect. 4.2.1 we have verified the leading logarithmic behaviour predicted by the resummation by performing a two-loop calculation in the soft and collinear limit. In order to do that we have considered the independent emission contribution to the soft eikonal current Eq. (4.24). For observables that are sensitive to emissions in the whole phase-space, such as for instance event shapes like thrust [86] a similar exercise can be also done for the correlated emission contribution to the soft current. Then we would find that these effects are fully accounted for by treating the running coupling in the CMW scheme, i.e. by considering the two-loop contribution to the cusp anomalous dimensions. However, it turns out that for so called non-global observables, i.e. observables that are sensitive only to a restricted region of phase-space, the all-order calculation previously described is not enough to capture full NLL accuracy. Indeed, correlated gluon emissions generate a new tower of single-logarithmic corrections [83, 87] the resummation of which is far from trivial.

Let us focus our discussion on a fixed-order example, which illustrates how a single logarithmic contribution arises in non-global observables. Because we are dealing with an observable that is only sensitive to emissions in a patch of the phase-space, we can have a configuration where a gluon is emitted outside this patch, in this case outside the jet, and it re-emits a softer gluon inside the jet. Thus, we consider the correlated emission contribution to the matrix element square for the emission of two soft gluons in the kinematic region where the harder gluon k_3 is not recombined with the jet, while the softer gluon k_4 is. In order to better illustrate the features of the calculation, in this section we are going to retain the full angular dependence, without taking the collinear limit. This makes sense because one of the gluons is emitted outside the jet, where the collinear approximation is less justified. Note that the integration over the gluon momentum k_3 is sensitive to the rest of the event and it may depend, for instance, on the way we select the jet, the mass of which we are measuring. For example, if we only select the hardest jet in the event, then one would have to prevent k_3 from clustering with k_2. For simplicity, in this example, we are going to integrate k_3 over the whole phase-space outside the measured jet. If we restrict ourselves to a jet algorithm, such as anti-k_t, which works as a perfect cone in the soft limit, this condition simply translates to $1 - \cos\theta_3 > 1 - \cos R$ and $1 - \cos\theta_4 < 1 - \cos R$. This situation is depicted in Fig. 4.3. At order α_s^2, the leading non-global contribution can be written as

$$\alpha_s^2 S^{(2)} = -4C_F C_A \left(\frac{\alpha_s}{2\pi}\right)^2 \int \frac{d\omega_3}{\omega_3} \int \frac{d\omega_4}{\omega_4} \Theta(\omega_3 > \omega_4) \int d\cos\theta_3$$

$$\times \int d\cos\theta_4 \, \Omega(\theta_3, \theta_4)$$

$$\Theta(\cos\theta_3 < \cos R) \, \Theta(\cos\theta_4 > \cos R) \, \Theta\left(\omega_4 Q(1 - \cos\theta_4) > m^2\right),$$

$$(4.34)$$

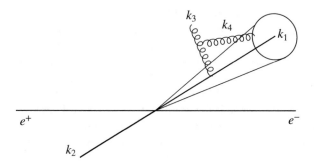

Fig. 4.3 Kinematic configuration that gives rise to non-global logarithms to lowest order in perturbation theory. The k_3 gluon is in the jet and does not contribute to the jet mass, while the k_4 gluon is in the jet and thus contributes to the jet mass

In this expression, the last Θ constraint comes from adding the real and virtual contributions for the gluon k_4. The angular function Ω arises after integrating the correlated matrix element square, Eq. (4.26), over the azimuth ϕ. Its expression reads [83]

$$\Omega(\theta_3, \theta_4) = \frac{2}{(\cos\theta_4 - \cos\theta_3)(1 - \cos\theta_3)(1 + \cos\theta_4)}. \tag{4.35}$$

We first perform the integration over the energies of the two gluons, obtaining

$$\alpha_s^2 S^{(2)} = -2C_F C_A \left(\frac{\alpha_s}{2\pi}\right)^2$$

$$\times \int d\cos\theta_3 \int d\cos\theta_4 \, \Theta(\cos\theta_3 < \cos R) \, \Theta(\cos\theta_4 > \cos R)$$

$$\Omega(\theta_3, \theta_4) \log^2\left(\frac{2m^2}{Q^2(1 - \cos\theta_4)}\right) \Theta\left(\frac{2m^2}{Q^2(1 - \cos\theta_4)} > 1\right). \tag{4.36}$$

We can now perform the angular integrations and express the results in terms of our rescaled variable ρ. The calculation can be simplified by noting that, since we are interested only in the NLL contribution, we can safely ignore the angular dependence in the argument of the logarithm. We obtain:

$$\alpha_s S^{(2)} = -2C_F C_A \left(\frac{\alpha_s}{2\pi}\right)^2 \frac{\pi^2}{6} \log^2\left(\frac{1}{\rho}\right) + \dots \tag{4.37}$$

where the dots indicate subleading contributions. It is interesting to observe that the coefficient of the first non-global logarithm is independent of R.[6] This might seem counter-intuitive at first because we might naively think that the probability for k_3 to emit a softer gluon inside the jet must be proportional to the jet area. However, the calculation shows that there is a nontrivial and R-independent contribution arising from the region where both gluons are close to the jet boundary. This results in an integrable singularity which is the origin of the $\pi^2/6$ contribution.

The result in Eq. (4.37) represents only the leading term at the first order at which non-global logarithms appear. In order to achieve full NLL accuracy these contributions must be resummed to all-orders. This is highly non-trivial, even if our aim is to resum only the leading tower of non-global logarithms needed at NLL. In order to perform an all-order analysis of non-global logarithms, we must consider configurations of many soft gluons. If we restrict ourselves to considering their leading contributions, which is single-logarithmic, we can assume energy-ordering; however, no collinear approximation can be made. Thus, we have to describe how an ensemble of an arbitrary number of soft gluons, all outside the jet, can emit an even softer gluon inside the jet.

Colour correlations make the colour algebra very complex as every emission increases the dimensionality of the relevant colour space. Moreover, describing the geometry of such ensembles also becomes difficult. The approach that was taken in the first analysis of non-global logarithms [83] was to consider the large-N_C limit. Colour correlations becomes trivial in this limit because the off-diagonal entries of the colour matrices vanish. Thus, we are able to write the matrix element square for the n gluon ensemble in a factorised way [88] and a simplified physical picture emerges. An emission off an ensemble of $n - 1$ gluons (plus the two hard partons) reduces to the sum over the emission off each of the n dipoles. When the dipole radiates a gluon, it splits into two dipoles, originating configurations which are determined by the history of the gluon branching. This can be implemented as a Monte Carlo which enables one to deal numerically with the second above-mentioned difficulty, namely the complicated geometry of the multi-gluon final states. This solution was first implemented in Ref. [83] and subsequently used in a number of phenomenological applications, e.g. [89–93].

The numerical impact of non-global logarithm on jet mass spectra can be large, see e.g. [92, 93], and because their treatment at NLL is only approximate, they often represent the bottleneck to reach perturbative precision in this kind of calculations. Remarkably, as we will discuss in Chap. 6, some grooming algorithms greatly reduce or even get rid of non-global logarithms, thus paving the way towards an improved perturbative accuracy of jet mass distributions.

Because of their complexity, a lot of effort has been invested in better understanding and controlling non-global logarithms. In the rest of this section, we highlight

---

[6]This result depends on the fact that we have integrated k_3 over the whole phase-space outside the jet. With additional constraints on the external region, the coefficient of $\log^2(\rho)$ would be more complex. However, in the small-R limit, one would always obtain $\frac{\pi^2}{6}$ up to powers of R.

some of the main results for the reader interested in a deeper exploration of non-global logarithms. The resummation of non-global logarithms was formalised by Banfi, Marchesini and Smye. In Ref. [94], they were able to derive an evolution equation, henceforth the BMS equation, which, equivalently to the Monte Carlo approach, resums the leading non-global logarithm, in the large-N_C limit. It has been noted [95] that the BMS equation has the same form as the Baliksty-Kochegov (BK) equation [96, 97] that describes non-linear small-x evolution in the saturation regime. This correspondence has been studied in detail in Refs [98, 99], where BMS and BK were related via a stereographic projection. Because a generalisation of the BK equation to finite N_C exists [100, 101], the correspondence between non-global logarithms and small-x physics was argued to hold at finite-N_C and numerical solutions have been studied [102, 103]. Very recently, this correspondence was indeed mathematically established [104]. In this approach, a *colour density matrix* is introduced, with the aim of describing soft radiation and an evolution equation is then derived for the colour density matrix, to all-loops, at finite N_C. The related anomalous dimension K is explicitly computed to one and two loops. The one-loop approximation to this evolution equation coincides with the BMS equation, once the large-N_C limit is taken and it confirms on a firmer ground the results of Refs. [102, 103] at finite N_C. More importantly, the explicit calculation of the two-loop contribution to K paves the way for the resummation of non-global logarithms at higher-logarithmic accuracy, although computing solutions to the evolution equation remains a challenging task.

A different approach to the question of resumming non-global logarithms was developed in Refs. [24, 25, 105] and applied to a phenomenological study of jet vetoes between hard jets in Refs. [106, 107]. In that context, because colour-correlations were of primary interest, the large-N_C limit did not seem adequate. We finish this discussion pointing out that other approaches similar in spirit was recently developed using techniques of SCET [108–112].

4.2.3   Dependence on the Clustering Algorithm

In all the calculations performed thus far we have always treated the constraints originating from the jet algorithm in a rather simple way. Essentially, we have always drawn a hard cone of radius R centred on the hard parton and considered as clustered into the jet soft emissions laying within that cone. As already mentioned, this approach is justified if we are using the anti-k_t algorithm. However, the situation changes for other members of the generalised k_t family, such as the Cambridge/Aachen algorithm or the k_t algorithms. Indeed, these clustering algorithms have a distance measure which admits the possibility of two soft gluons being the closest pair, thus combining them before they cluster with the hard parton.

We now revisit the two-gluon calculation described in Sect. 4.2.1, this time making use of the k_t clustering algorithm. We keep the same convention for the kinematics, i.e. the soft gluon momenta are labelled k_3 and k_4, with k_4 much softer than k_3. As in the previous section, we should consider either the case where both gluons are real, or the case where one of the gluons (either k_3 or k_4) is real and the other is virtual.

We start by considering the RR contribution in different kinematic configurations. Clearly, when both k_3 and k_4 are beyond an angle R with respect to the hard parton there is no contribution from either to the jet-mass. When both k_3 and k_4 are within an angle R of the hard parton, both soft gluons get combined into the hard jet and this region produces precisely the same result as the anti-k_t algorithm, corresponding to exponentiation of the one-gluon result.[7] However, when k_3 is beyond an angle R and k_4 is inside an angle R the situation changes from the anti-k_t case. This is because the k_t distance between the two soft gluons can be smaller than the k_t distance between k_4 and the hard parton, in which case k_4 clusters with k_3, resulting in a soft jet along the direction of k_3. Thus, when k_3 is beyond an angle R it can pull k_4 out of the hard jet since the soft jet $k_3 + k_4$ lays now at angle larger than R with respect to the hard parton, i.e. outside the jet. Therefore, this kinematic configuration results in a massless jet. In precisely the same angular region the VR configuration is obviously unaffected by clustering and it does give a contribution to $\frac{d\sigma}{d\rho}$. This contrasts with the anti-k_t case where the real and virtual contributions cancelled exactly at this order. Note also that the RV configuration gives no contribution (as in the anti-k_t case) because no real gluons are in the jet. Finally, for the case were k_3 is inside the jet and k_4 is outside the jet, a similar situation can happen where k_3 and k_4 are clustered first, pulling k_4 back in the jet. This case however does not lead to an extra contribution because, since k_4 is much softer than k_3, it does not affect the mass of the jet already dominated by k_3.

Thus a new contribution arises for the k_t algorithm from the region where the two real gluons k_3 and k_4 are clustered, where we only get a contribution from the case where k_3 is virtual and k_4 is real. We now carry out this calculation explicitly. We work in the small-R limit and consider the angles θ_3, θ_4 and θ_{34} as the angles between k_3 and the hard parton, k_4 and the hard parton and k_3 and k_4 respectively. In order to apply the k_t-algorithm in e^+e^-, we have to compare the distances $\omega_3^2\theta_3^2$, $\omega_4^2\theta_4^2$ and $\omega_4^2\theta_{34}^2$. Now since $\theta_3^2 > R^2$, $\theta_4^2 < R^2$ and $\omega_4 \ll \omega_3$, the only quantities that can be a candidate for the smallest distance are $\omega_4^2\theta_4^2$ and $\omega_4^2\theta_{34}^2$. Thus the gluons are clustered and k_4 is pulled out of the jet if $\theta_{34} < \theta_4 < R$. Otherwise k_4 is in the jet and cancels against virtual corrections, precisely as it happened for the anti-k_t algorithm.

---

[7]Remember that when a soft particle clusters with a much harder one, the resulting object has the p_t and direction of the harder particle, up to negligible recoil.

Making use of the usual rescaling $\theta \rightarrow \theta/R$, we can then write the VR contribution in the clustering region as

$$
\frac{d\tilde{\sigma}_2^{\text{cluster}}}{d\rho} = -4C_F^2 \left(\frac{\alpha_s}{2\pi}\right)^2 \int \frac{d\theta_3^2}{\theta_3^2} \frac{d\theta_4^2}{\theta_4^2} \frac{d\phi}{2\pi} \frac{dz_3}{z_3} \frac{dz_4}{z_4} \delta\left(\rho - z_4\theta_4^2\right) \Theta(z_3 > z_4)
$$
$$
\Theta\left(\theta_3^2 > 1\right) \Theta\left(\theta_{34}^2 < \theta_4^2\right) \Theta\left(\theta_2^2 < 1\right). \qquad (4.38)
$$

Within our small-angle approximation, we can write $\theta_{34}^2 = \theta_3^2 + \theta_4^2 - 2\theta_3\theta_4 \cos\phi$. Integrating over z_3 and z_4 and using $t = \frac{\theta_4^2}{\rho}$ one obtains

$$
\frac{d\tilde{\sigma}_2^{\text{cluster}}}{d\rho} = -4C_F^2 \left(\frac{\alpha_s}{2\pi}\right)^2 \frac{1}{\rho} \int \frac{d\theta_3^2}{\theta_3^2} \frac{dt}{t} \frac{d\phi}{2\pi} \log(t)
$$
$$
\Theta\left(t > 1\right) \Theta\left(\theta_3^2 > 1\right) \Theta\left(4\rho t \cos^2\phi > \theta_3^2\right) \Theta\left(t\rho < 1\right). \qquad (4.39)
$$

Carrying out the integral over θ_3^2 results in

$$
\frac{d\tilde{\sigma}_2^{\text{cluster}}}{d\rho} = -4C_F^2 \left(\frac{\alpha_s}{2\pi}\right)^2 \frac{1}{\rho} \int \frac{dt}{t} \frac{d\phi}{2\pi} \log\left(4\rho t \cos^2\phi\right) \log(t)
$$
$$
\Theta\left(t > 1\right) \Theta\left(4\rho t \cos^2\phi > 1\right) \Theta\left(\rho t < 1\right). \qquad (4.40)
$$

Now we need to carry out the t integral for which we note $t > \max\left(1, \frac{1}{4\rho\cos^2\phi}\right)$. In the region of large logarithms which we resum one has however that $\rho \ll 1$ and hence $4\rho\cos^2\phi \ll 1$. At NLL accuracy we can therefore take $t > \frac{1}{4\rho\cos^2\phi}$ and replace the $\log(t)$ factor by $\log\left(\frac{1}{\rho}\right)$ in (4.40). It is then straightforward to carry out the t integration to get

$$
\frac{d\tilde{\sigma}_2^{\text{cluster}}}{d\rho} = -4C_F^2 \left(\frac{\alpha_s}{2\pi}\right)^2 \frac{1}{\rho} \log\left(\frac{1}{\rho}\right) \int_{-\frac{\pi}{3}}^{\frac{\pi}{3}} \frac{d\phi}{\pi} \log^2(2\cos\phi)
$$
$$
= -\frac{2\pi^2}{27} C_F^2 \left(\frac{\alpha_s}{2\pi}\right)^2 \frac{1}{\rho} \log\left(\frac{1}{\rho}\right). \qquad (4.41)
$$

This behaviour in the distribution corresponds to a single-logarithmic $\alpha_s^2 \log^2\left(\frac{1}{\rho}\right)$ contribution to the cumulative, which is, as anticipated, necessary to claim NLL accuracy. The all-order treatment of these clustering effects is far from trivial because of the complicated kinematic configurations, which results into many nested Θ function. Therefore, from this point of view, resummation of mass spectra for jet defined with the anti-k_t algorithm appears simpler. Conversely, because of these clustering effects, the jet boundary becomes somewhat blurred, resulting in milder non-global contributions.

4.2.4   Non-perturbative Corrections: Hadronisation

Lund diagrams, such as the one in Fig. 4.1, turn out to be particularly useful in order to determine the sensitivity of an observable to non-perturbative dynamics. We can introduce a non-perturbative scale $\mu_{NP} \sim 1$ GeV below which we enter a non-perturbative regime. Because the running coupling in Eq. (4.14) is evaluated at a scale that represent the emission transverse momentum with respect to the jet, a horizontal line $z\theta = \tilde{\mu} = \frac{\mu_{NP}}{E_J R}$ marks the boundary between perturbative and non-perturbative dynamics (recall that θ is measured in unit of the jet radius R). It is then simple to calculate what is the corresponding value of the jet mass for which the integrals we have to perform have support on the non-perturbative region: we just have to work out where the line of constant ρ first crosses into the non-perturbative region. This happens when $z\theta = \tilde{\mu}$ and $\theta = 1$, which implies $\rho = \tilde{\mu}$. Thus, this simple argument suggests that the mass distribution becomes sensitive to non-perturbative physics at

$$m^2 \simeq \frac{\mu_{NP}}{E_J R} E_J^2 R^2 = \mu_{NP} E_J R. \tag{4.42}$$

Note that this scale grows with the jet energy, so that even apparently large masses, $m \gg \Lambda_{QCD}$, may in fact be driven by non-perturbative physics. For a 3 TeV jet with $R = 1$, taking $\mu_{NP} = 1$ GeV, the non-perturbative region corresponds to $m \lesssim 55$ GeV, disturbingly close to the electroweak scale!

Experimentally jets can be thought of as a bunch of collimated hadrons (mesons and baryons). However, we have so far considered jets from a perturbative QCD perspective and used partons to describe their constituents. The parton-to-hadron transition, namely hadronisation, is a non-perturbative phenomenon. Non-perturbative corrections due to hadronisation can be treated, within certain approximations, with analytic methods, see e.g. [113, 114]. For the jet mass the leading correction turns out to be a shift of the differential distribution [115, 116]. Furthermore, this type of analytic calculations can provide insights about the dependence of these corrections on the parameters of the jet algorithm, such as the jet radius [79]. Alternatively, we can take a more phenomenological point of view and use Monte Carlo parton showers to estimate non-perturbative correction. For instance, we can either calculate a given observable on a simulated event with hadrons in the final state, or stop the event simulation before hadronisation takes place and compute the same observable with partons. We can then take the bin-by-bin ratio of the jet mass distribution computed with and without hadronisation as a proxy for these corrections. This is the path we are going to employ in this book to illustrate the impact of non-perturbative corrections (both hadronisation and the Underlying Event, which we must also include when considering hadron-hadron collisions). We will present such studies in Chap. 6, where hadronisation correction to the jet mass distribution discussed here will be compared to the ones for jets with substructure (typically grooming) algorithms.

4.3 From e^+e^- to Hadron-Hadron Collisions

Thus far we have discussed the resummation of the invariant mass distribution of a jet produced in an electron-positron collision. In order to be able to perform jet studies in proton-proton collision we have to extend the formalism developed so far. A detailed derivation of the resummation formulae goes beyond the scope of this book and we refer the interested reader to the original literature, e.g. [117, 118]. Here, instead we briefly sketch the issues that we have to tackle and how we can go about them.

(a) As discussed in Chap. 2, in proton-proton collision, we work in the collinear factorisation framework, Eq. (2.4), where cross-sections are described as a convolution between a partonic interaction and universal parton distribution functions. Furthermore, we need to switch to the appropriate kinematic variables for proton-proton collisions, namely transverse momentum, rapidity and azimuthal angle (cf. Sect. 2.5).

(b) The complexity of resummed calculations increases in the case of hadronic process because we have to deal with many hard legs with colour, including the initial-state partons. As we have noted in Sect. 2.3, factorisation in the soft limit happens at the amplitude level and interference terms play a crucial role in the soft limit.

As a consequence, resummed calculations that aim to correctly capture these effect must account for all non-trivial colour configuration. In particular, if we have a process that at Born level as more than two coloured hard legs, either in the initial or final state, then the one-gluon emission contribution in the soft limit Eq. (4.3) can be generalised as follows

$$W = \sum_{(ij)} \frac{\alpha_s(\kappa_{ij}) \, C_{ij}}{2\pi} \frac{p_i \cdot p_j}{(p_i \cdot k)(p_j \cdot k)}, \qquad (4.43)$$

where to avoid confusion we have labelled the momenta of the hard legs as p_i (rather than k_i) and k is the soft gluon momentum. We note that the sum runs all over the dipoles (ij), i.e. all pairs of hard legs i and j. To NLL accuracy, the running coupling in Eq. (4.43) must be evaluated at the scale $\kappa_{ij}^2 = \frac{2(p_i \cdot k)(p_j \cdot k)}{(p_i \cdot p_j)}$, which is the transverse momentum of the emission with respect to the dipole axis, in the dipole rest frame. C_{ij} is a generalisation of the effective colour charge, Eq. (2.16), which is not necessarily diagonal:

$$C_{ij} = -2\, T_i \cdot T_j, \tag{4.44}$$

where the colour matrices T_i are not necessarily in the fundamental representation, as the gluon can be emitted off a gluon line as well. We note that the expression above greatly simplifies in the collinear limit, where one recovers the usual colour factors C_F and C_A. However, soft emissions at large angle do contribute beyond LL and therefore dealing with the sum over dipoles is mandatory in order to achieve NLL accuracy.

It is possible to show that, even in the presence of many hard legs, the one-loop contribution above still exponentiates. However, one must keep track, for each dipole, of the different colour flow configurations. This results into a rather complex matrix structure in colour space [117, 118]. As an example, in Sect. 4.3.1, we will evaluate the contribution to the jet mass distribution in pp collision from a soft gluon emission emitted from the dipole made up of the incoming hard legs.

(c) Finally, new sources of non-perturbative corrections arise in proton-proton collisions. Collinear factorisation assumes that only one parton from each proton undergoes a hard scattering. However, we can clearly have secondary, softer, scatterings between the protons' constituents. As we have mentioned at the beginning of this book, these multiple-parton interactions produce what is usually referred to as the Underlying Event. Furthermore, because protons are accelerated and collided in bunches, we also have multiple proton-proton interactions per bunch-crossing, leading to what we call pileup. As a consequence hadronic collisions are polluted by radiation that does not originated from the hard scattering. In the context of jet physics this radiation has important consequences as it modifies the jet properties, e.g. its transverse momentum or its mass, in a way which is proportional to powers the jet radius R. More specifically, corrections to the jet transverse momentum are proportional to R^2, while corrections to the jet mass exhibit a R^4 behaviour [79]. Therefore large-R jets are more significantly affected by these effects.

Some first-principle studies have been performed, mostly concentrating on double-parton scattering (see Ref. [119] for a recent review), however most phenomenological analyses rely on models of the underlying event which are usually incorporated in Monte Carlo simulations. These models are characterised by a number of free parameters which are determined by comparisons with experimental data with a process known as *tuning*. We will come back to the numerical impact of the underlying event in Chap. 6, where we will discuss the ability of grooming techniques to reduce such contamination.

To illustrate the extra complications one has to deal with in proton-proton collisions, we conclude this chapter by computing first the effect of initial-state radiation and then the jet mass distribution in Z+jet events.

4.3.1   Initial-State Radiation as an Example

In this section we sketch the calculation of the contribution to the jet mass distribution from the emission of a soft a gluon from the dipole formed by the two incoming hard legs. This can be taken as a good proxy to the effect of initial-state radiation. As it is the first calculation we perform with hadron-collider kinematic variables, let us explicitly specify the kinematics:

$$p_1 = \frac{\sqrt{s}}{2} x_1 \, (1, 0, 0, 1) \,,$$

$$p_2 = \frac{\sqrt{s}}{2} x_2 \, (1, 0, 0, -1) \,,$$

$$p_3 = p_t \, (\cosh y, 1, 0, \sinh y) \,,$$

$$k = k_t \, (\cosh \eta, \cos \phi, \sin \phi, \sinh \eta) \,, \tag{4.45}$$

where p_1 and p_2 denote the four-momenta of the incoming hard partons, p_3 the momentum of the jet, and k of the soft gluon. It is understood that the jet must recoil against a system with momentum p_4 (not specified above), over which we are inclusive. Note that we have used hadron-collider variables, i.e. transverse momenta p_t and k_t, rapidities y and η, and azimuthal angle ϕ, assuming without loss of generality that the jet is produced at $\phi = 0$. Provided the soft gluon is clustered with the jet, its contribution to the jet mass is

$$m^2 = (p_3 + k)^2 = 2 p_3 \cdot k = 2 p_t k_t \, (\cosh(\eta - y) - \cos \phi) \,. \tag{4.46}$$

We can now write the contribution to the cumulative distribution from the 12 as

$$\alpha_s \Sigma_{12}^{(1)} = C_{12} \int k_t dk_t d\eta \frac{d\phi}{2\pi} \frac{\alpha_s (\kappa_{12})}{2\pi} \frac{(p_1.p_2)}{(p_1.k)(p_2.k)} \Theta \left((\eta - y)^2 + \phi^2 < R^2 \right)$$

$$\cdot \, \Theta \left(\frac{2 k_t}{p_t R^2} (\cosh(\eta - y) - \cos \phi) > \rho \right) \,, \tag{4.47}$$

where the first Θ function is the jet clustering condition and we have introduced $\rho = \frac{m^2}{p_t^2 R^2}$, analogously to the e^+e^- case. We next note that

$$\kappa_{12}^2 = 2 \frac{(p_1.k)(p_2.k)}{(p_1.p_2)} = k_t^2 \,. \tag{4.48}$$

Equation (4.47) therefore exhibits a logarithmic enhancement at small k_t as expected. To isolate the leading (NLL) contribution, we can as usual just retain the dependence of the jet mass on k_t in the second line of (4.47), and neglect the dependence on y, η and ϕ which produces terms beyond NLL accuracy. We can then carry out the integration over η and ϕ which simply measures the jet area πR^2 and obtain

$$\alpha_s \Sigma_{12}^{(1)} = C_{12} R^2 \int_{\rho p_t}^{p_t} \frac{\alpha_s(k_t)}{2\pi} \frac{dk_t}{k_t}, \qquad (4.49)$$

where the lower limit of integration stems from the constraint on the jet mass. The dipole consisting of the two incoming partons gives indeed rise to a pure single-logarithmic behaviour. Since the emitted gluon is inside the jet region, away from the hard legs constituting the dipole, there are no collinear enhancements. Furthermore, the soft wide-angle single logarithm we obtain is accompanied by an R^2 dependence on jet radius, reflecting the integration over the jet area.

4.3.2 The Jet Mass Distribution in $pp \to Z$+jet

We finish this chapter by showing how all the effects discussed so far affects the calculation of a jet mass distribution. We choose to study the jet mass distribution of the hardest jet produced in association with a Z boson. This process is of particular interest in the boosted regime $p_t \gg m$ or, equivalently, $\rho \ll 1$ because it is the main background for the production of a boosted Higgs boson, recoiling against the Z. In practice, it also has a simpler structure than the jet mass in dijet events since there are only three coloured hard legs.

At Born level we have to consider two partonic processes $qg \to Zq$ and $q\bar{q} \to Zg$. We can think as the first process to describe the production of a quark-initiated jet, while the second one gives a gluon-initiated jet. We consider a very hard jet with $p_t = 3$ TeV and jet radius $R = 1$.

We plot in Fig. 4.4 the distribution of the variable ρ calculated to NLL in several approximations, on the left for a quark-initiated jet, and on the right for a gluon-initiated one. We start by considering the exponentiation of the single gluon emission Eq. (4.17), in the collinear, i.e. small R limit (dotted red curve). We then add the contribution to multiple emission Eq. (4.22) (dash-dotted green curve). We then add the correction due to non-global logarithms in the large-N_C limit [83] (dashed blue curve). Finally, we include corrections which are suppressed by powers of the jet radius (solid black curve).

To illustrate basic aspects of the colour algebra, we work out the effective colour factors C_{ij} associated to the colour dipoles (cf. Eq. (4.44)) of our Z+jet process. Let us start with the $qg \to Zq$ process and label with 1 the incoming quark, with 2 the incoming gluon and with 3 the outgoing quark, we have that $T_1^2 = T_3^2 = C_F$ and $T_2^2 = C_A$. Exploiting colour conservation, i.e. $T_1 + T_2 + T_3 = 0$ (with all dipole legs considered outgoing), we find

$$C_{12} = C_{23} = C_A = N_C, \qquad C_{13} = 2C_F - C_A = -\frac{1}{N_C}. \qquad (4.50)$$

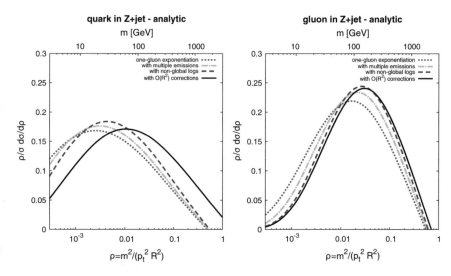

Fig. 4.4 The mass distribution of the quark-initiated and gluon-initiated jets in Z+jet. The numerical impact of different contributions at NLL accuracy is shown

We then move to the gluon-initiated jet case, i.e. the Born process $q\bar{q} \to Zg$ and label with 1 the incoming quark, with 2 the incoming antiquark and with 3 the outgoing gluon. We have that $T_1^2 = T_2^2 = C_F$ and $T_3^2 = C_A$ and

$$C_{12} = 2C_F - C_A = -\frac{1}{N_C}, \qquad C_{13} = C_{23} = C_A = N_C. \qquad (4.51)$$

We note that the $\mathcal{O}(R^2)$ corrections are rather sizeable because we are dealing with a jet with large radius. However, further corrections $\mathcal{O}(R^4)$ turn out to be very small and indistinguishable on the plot. The bulk of this large $\mathcal{O}(R^2)$ effect originates from the 12 dipole studied above, i.e. it can be thought as the contribution of initial-state radiation to the jet mass. Finally, we remind the reader that the result in Fig. 4.4 is not matched to fixed-order and therefore it is not reliable in the $\rho \sim 1$ region. In particular, the resummation is not capable to correctly capture the end-point of the distribution and matching to (at least) NLO is mandatory to perform accurate phenomenology.

Chapter 5
Jet Substructure: Concepts and Tools

The widest application of jet substructure tools is to disentangle different kinds of jets. This typically includes separating quark and gluon-initiated jets or isolating boosted W/Z/H or top jets (our signal) from the much more abundant QCD background of "standard" quark and gluon-initiated jets. In this chapter, we discuss these methods in some detail. We start by considering the guiding principles behind the different algorithms and how to assess their performance. Then, we will review some of the most commonly-used jet substructure techniques over the past 10 years. Explicit examples on how these tools behave in Monte-Carlo simulations and analytic calculations and how they are used in experimental analyses will be given in the next Chapters.

5.1 General Guiding Principles

Jet substructure aims to study the internal kinematic properties of a high-p_t jet in order to distinguish whether it is more likely to be a signal or background jet. Although a large variety of methods have been proposed over the last 10 years, they can be grouped into three wide categories, according the physical observation that they mostly rely on.

Category I: prong finders. Tools in this category exploit the fact that when a boosted massive object decays into partons, all the partons typically carry a sizeable fraction of the initial jet transverse momentum, resulting in multiple hard cores in the jet. Conversely, quark and gluon jets are dominated by the radiation of soft gluons, and are therefore mainly single-core jets. Prong finders therefore look for multiple hard cores in a jet, hence reducing the contamination from "standard" QCD jets. This is often used to characterise the boosted jets in terms of their "pronginess", i.e. to their expected number of hard cores: QCD jets would be 1-prong objects, W/Z/H jets would be two-pronged, boosted top

© Springer Nature Switzerland AG 2019
S. Marzani et al., *Looking Inside Jets*, Lecture Notes in Physics 958,
https://doi.org/10.1007/978-3-030-15709-8_5

jets would be three-pronged, an elusive new resonance with a boosted decay into two Higgs bosons, both decaying to a $b\bar{b}$ pair would be a 4-prong object, . . .

Category II: radiation constraints.   The second main difference between signal and background jets is their colour structure. This means that signal and background jets will exhibit different soft-gluon radiation patterns. For example, QCD radiation associated with an EW-boson jet, which is colourless, it is expected to be less than what we typically find in a QCD jet. Similarly, quark-initiated jets are expected to radiate less soft gluons than gluon-initiated jets. Many jet shapes have been introduced to quantify the radiation inside a jet and hence separate signal jets from background jets.

Category III: groomers.   There is a third category of widely-used tools related to the fact that one often use large-radius jets for substructure studies. As we have already discussed, because of their large area, these jets are particularly sensitive to soft backgrounds, such as the UE and pileup. "Grooming" tools have therefore been introduced to mitigate the impact of these soft backgrounds on the fat jets. These tools usually work by removing the soft radiation far from the jet axis, where it is the most likely to come from a soft contamination rather than from QCD radiation inside the jet. In many respects, groomers share similarities with prong finders, essentially due to the fact that removing soft contamination and keeping the hard prongs are closely related.

Additionally, we note that we might expect non-trivial interplay between groomers and radiation-constraint observables. For instance, if we apply observables that exploit radiation constraints on soft radiation, to groomed jets, which precisely throws away soft radiation, we expect to obtain worse performance. Therefore, we can anticipate that we will have to find a sweet spot between keeping the sensitivity to UE and pileup under control, while maintaining a large discriminating power.

5.2 Assessing Performance

Even though they are based on only a handful of key concepts, a long list of jet substructure tools have been introduced. Before we dive into a description of these tools, it is helpful to briefly discuss how one can compare their relative performance. Note that, here, we are not referring to how the tools can be validated, which is often do via Monte Carlo studies, direct measurements in data or analytic studies. Instead, we would like to answer questions like "There are dozens of tools around, which one should I use for my problem?" or "Which one has the largest performance?". It is of course impossible to give a definite answer to such questions, but what we can at least provide is some key ideas of what we mean by "performant" which can be properly tested and quantified.

The case of groomers is the probably the easiest to address, since groomers have the specific purpose of suppressing the sensitivity to the UE and pileup. In the case of the UE, we can perform Monte Carlo studies, switching multiple particle interactions on and off, to check how key distributions—like the jet mass distribution in QCD events, W→ $q\bar{q}$ or hadronic top decay—vary. A similar approach applies for pileup, where we can perform Monte Carlo studies, overlaying minimum bias events with the hard events. In cases where we have access to both a reference event (e.g. a hard collision) and a modified event (e.g. the same event overlaid with pileup), quality measures can then involve average shifts and dispersions of how jet quantities like the jet mass are affected event by event. More generally, we can study the position and width of peaks like the reconstructed W or top mass, and study their stability with respect to the UE or pileup multiplicity. We refer to Section 4 of Ref. [8] for an explicit application of the above procedure.

In the following, we are going to focus on the case of boosted-object tagging. In this case, there is again a very obvious meaning of what performant means: the best tool is the one which keeps most of the signal and rejects most of the background. In practice, for a signal S and a background B, we define the signal (respectively background) efficiency ϵ_S (ϵ_B) as the rate of signal (or background) jets that are accepted by the tagger. For cases with limited statistics (which is often the case in searches), the best tool is then the one that maximises the signal significance, $\epsilon_S/\sqrt{\epsilon_B}$. More generally, for a given signal efficiency, one would like to have the smallest possible background rate, i.e. for a given amount of signal kept by the tagger, we want to minimise the rate of background events which wrongly pass the tagger conditions. This is usually represented by Receiver Operating Characteristic (ROC) curves which show ϵ_B as a function of ϵ_S, such as represented in Fig. 5.1. This can be used to directly compare the performance of different substructure tools.

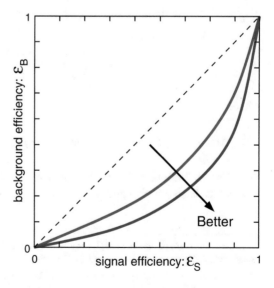

Fig. 5.1 A ROC curve represents the background efficiency ϵ_B as a function of the signal efficiency ϵ_S. For a given signal efficiency, a lower background efficiency is better

That said, signal significance is not the only criterion one may desire from a jet substructure tagger. Similarly to the properties of jet definitions discussed in Chap. 3, we may want additional conditions such as the following:

- we would like to work with tools that are infrared and collinear safe, i.e. which are finite at any order of the perturbation theory,[1]
- we would like to work with tools that are as little sensitive as possible to model-dependent non-perturbative effects such as hadronisation and the Underlying Event,
- we would like to work with tools that are as little sensitive as possible to detector effects and pileup.

In a way, the last two of the above criteria are related to the *robustness* of our tools, i.e. we want to be able to assess how robust our conclusions are against details of the more poorly-known (compared to the perturbative part) aspects of high-energy collisions. One should typically expect that a more robust tool would have a smaller systematic uncertainty associated with theory modelling (e.g. the dependence on which Monte Carlo sample is used), pileup sensitivity and detector sensitivity/unfolding.[2]

Robustness can be quantified in several ways, typically by measuring how the signal and background efficiencies are affected by a given effect (see e.g. [120–122]). Some concrete ideas about how to assess robustness were put forward in Ref. [122] (Section III.2). Let us say that we want to test the sensitivity of a tagger with respect to the UE. From a Monte Carlo simulation, we can compute the signal and background efficiencies, first without UE, $\epsilon_{S,B} \equiv \epsilon_{S,B}^{(\text{no UE})}$, and then with UE $\epsilon'_{S,B} \equiv \epsilon_{S,B}^{(\text{UE})}$. We define *resilience*, a measure of robustness, as

$$\zeta = \left(\frac{\Delta \epsilon_S^2}{\langle \epsilon \rangle_S^2} + \frac{\Delta \epsilon_B^2}{\langle \epsilon \rangle_B^2} \right)^{-1/2} \tag{5.1}$$

where

$$\Delta \epsilon_{S,B} = \epsilon_{S,B} - \epsilon'_{S,B} \quad \text{and} \quad \langle \epsilon \rangle_{S,B} = \frac{1}{2} \left(\epsilon_{S,B} + \epsilon'_{S,B} \right). \tag{5.2}$$

With this definition, a large resilience means that the signal and background efficiencies have not changed much when switching the UE on and hence that the

---

[1] An interesting class of observables, known as *Sudakov safe*, fails to fully satisfy this condition but remain calculable once a proper all-order calculation is performed (see Chap. 9).

[2] Small systematic uncertainties is really the fundamental assessment of robustness. Asking, as we do here, for a small sensitivity to non-perturbative and detector effects, is a sufficient condition to achieve this, but it is not strictly necessary. One could for example imagine a situation where detector effects are large but perfectly well understood such that the resulting systematic uncertainty remains small.

tool is robust. Resilience can be defined for hadronisation, i.e. when switching on hadronisation and going from parton level to hadron level, for the UE, as discussed above, for pileup sensitivity, i.e. when overlaying the event with pileup and applying a pileup mitigation technique, and for detector sensitivity, i.e. when running events through a detector simulation.

To conclude, it is important to realise that the performance of a jet substructure tagger is characterised by several aspects. Performance, typically quantified by ROC curves of signal significance is certainly the most regarded feature of a tagger. However, other requirements like the robustness against non-perturbative effects, pileup and detector effects are desirable as well. These can be quantified e.g. via resilience.

5.3   Prong-Finders and Groomers

Mass-Drop Tagger   The Mass-Drop tagger was originally proposed [5] as a tool to isolate boosted Higgs bosons, decaying to $b\bar{b}$ pairs, from the QCD background. In this procedure, one first reclusters the jet constituents of the fat jet with the Cambridge/Aachen algorithm. One then iteratively undoes the last step of the clustering $p_{i+j} \rightarrow p_i + p_j$ and check the following criteria: (1) there is a "mass drop" i.e. $\max(m_i, m_j) < \mu_{\text{cut}} m_{i+j}$, (2) the splitting is sufficiently symmetric i.e. $\min(p_{t,i}^2, p_{t,j}^2) \Delta R_{ij}^2 > y_{\text{cut}} m_{i+j}^2$. When both criteria are met, we keep "$i + j$" as the result of the mass-drop tagger, otherwise the least massive of i and j is discarded and the procedure is repeated iteratively using the most massive of i and j.[3] The mass-drop tagger has two parameters: μ_{cut}, the mass-drop parameter itself, and y_{cut}, the symmetry cut. The two conditions imposed by the mass-drop tagger exploit the fundamental properties introduced above for tagging two-pronged boosted objects: the symmetry cut requires that one indeed finds two hard prongs and the mass-drop condition imposes that one goes from a massive boson jet to two jets originated from massless QCD partons. Although it was originally introduced as a tagger, the mass-drop tagger also acts as a groomer since, following the declustering procedure, it would iteratively remove soft radiation at the outskirts of the jet, hence reducing the pileup/UE contamination.

modified Mass-Drop Tagger (mMDT)   When trying to understand the analytic behaviour of the mass-drop tagger on QCD jets, it was realised that following the most massive branch in the iterative de-clustering procedure leads to pathological situations. It was therefore suggested [123] to adapt the procedure so that it instead follows the hardest branch (in terms of p_t). This modification makes the analytical calculation much easier and more robust without affecting the performance of

---

[3]If the procedure fails to find two subjets satisfying the conditions, i.e. end up recursing until it reaches a single constituent which can not be further de-clustered, it is considered as having failed and returns an empty jet.

the method (even improving it slightly). The same study also added two more minor modifications. First, it was realised that the symmetry condition could be replaced by $\min(p_{t,i}, p_{t,j}) > z_{\text{cut}}(p_{t,i} + p_{t,j})$ which has the same leading analytic behaviour as the y_{cut} condition and a slightly reduced sensitivity to non-perturbative corrections. Second, the mass-drop condition would only enter as a subleading correction in the strong coupling constant α_s, compared to the symmetry condition. It can therefore usually be ignored.

SoftDrop SoftDrop [50] can be seen a generalisation of mMDT. It also proceeds by iteratively declustering a jet reclustered with the Cambridge/Aachen algorithm but replaces the symmetry condition for the declustering of p_{i+j} into p_i and p_j, with

$$\frac{\min(p_{t,i}, p_{t,j})}{p_{t,i} + p_{t,j}} > z_{\text{cut}} \left(\frac{\Delta R_{ij}}{R} \right)^{\beta}, \tag{5.3}$$

where R is the jet radius. SoftDrop has two parameters. The z_{cut} parameter plays the same role as in the (m)MDT of keeping the hard structure and excluding soft emissions, starting from large angles. The β parameter gives SoftDrop some extra freedom in controlling how aggressive the groomer is. In the limit $\beta \to 0$, SoftDrop reduces to the mMDT. Increasing β leads to a less aggressive grooming procedure, with $\beta \to \infty$ corresponding to no grooming at all. Conversely, choosing a negative value for β would lead a more aggressive two-prong tagger than mMDT.[4] For practical applications, mMDT and SoftDrop with negative β (typically $\beta = -1$) would, alone, be perfectly adequate and efficient taggers (see e.g. Section 7 of Ref. [50]).

Recursive SoftDrop SoftDrop typically finds two prongs in a jet. If we want to find more than two prongs, we can apply SoftDrop recursively. Recursive SoftDrop [124] does this by iteratively undoing the clustering with the largest ΔR in the Cambridge/Aachen tree. Both branches are kept if the SoftDrop condition (5.3) is met and the softer branch is dropped otherwise. The procedure stops when $N + 1$ prongs have been found, with N an adjustable parameter that can be taken to infinity.

Filtering Filtering was first introduced in Ref. [5] as a grooming strategy to clean the jet from UE after the mMDT has been applied. For a given jet, it re-clusters its constituents with the Cambridge/Aachen algorithm with a small radius R_{filt} and only keeps the n_{filt} larger p_t subjets. The subjets that have been kept constitute filtered jet. This has two adjustable parameters: R_{filt} and n_{filt}. It is typically used to reduce soft contamination in situations where we have a prior knowledge of the number of hard prongs in a jet. For a jet with n_{prong} hard prongs—$n_{\text{prong}} = 2$ for a $W/Z/H$

---

[4]The SoftDrop procedure returns by default a single particle if it fails to find two subjets satisfying the SoftDrop condition. This "grooming mode" is different from the default "tagging mode" of the mMDT which would fail, i.e. return an empty jet, if no substructure are found.

bosons and $n_{prong} = 3$ for a top—we would typically use $n_{filt} = n_{prong} + 1$ which would also keep the (perturbative) radiation of an extra gluon.

Trimming Trimming [125] shares some similarities with filtering. It also starts with re-clustering the jet with a smaller radius, R_{trim}, using either the k_t or the Cambridge/Aachen algorithm. It then keeps all the subjets with a transverse momentum larger than a fraction f_{trim} of the initial jet transverse momentum. On top of the choice of algorithm, this also has two parameters: R_{trim} and f_{trim}. It is often used both as a generic groomer and as a prong finder in boosted-jet studies.

Pruning Pruning [126] is similar in spirit to trimming but it adopts a bottom-up approach (with trimming seen as a top-down approach). Given a jet, pruning reclusters its constituents using a user-specified jet definition (based on pairwise recombinations) and imposes a constraint at each step of the clustering: objects i and j are recombined if they satisfy at least one of these two criteria: (1) the geometric distance ΔR_{ij} is smaller than $R_{prune} = 2 f_{prune} m_{jet}/p_{t,jet}$, with $p_{t,jet}$ and m_{jet} the original jet transverse momentum and jet mass, (2) the splitting between i and j is sufficiently symmetric, i.e. $\min(p_{t,i}, p_{t,j}) \geq z_{prune} p_{t,(i+j)}$. If neither criteria are met, only the hardest of i and j (in terms of their p_t) is kept for the rest of the clustering and the other is rejected. On top of the jet definition used for the re-clustering, which is usually taken to be either k_t or Cambridge/Aachen with a radius much larger than the one of original jet, this has two parameters: f_{prune} and z_{prune}. z_{prune} plays the same role as f_{trim} for trimming and f_{prune} plays a role similar to R_{trim}. Note that, in the case of pruning, R_{prune} is defined dynamically based on the jet kinematics, while R_{trim} is kept fixed. This can have important consequences both analytically and phenomenologically. Pruning can be considered as a general-purpose groomer and tagger and is often used in situations similar to trimming, although it tends to be slightly more sensitive to pileup contamination.

I and Y-Pruning When pruning a jet, there might be situations where a soft emission at large angle dominates the mass of the jet, thus setting the pruning radius, but gets pruned away because it does not satisfy the pruning conditions. The mass of the pruned jet is then determined by radiation at smaller angle, typically within the pruning radius. This situation where the jet mass and the pruning radius are determined by different emissions in the jet would result in a jet with a single prong, and it usually referred to called "I-pruning" [123]. For I-pruning, the pruning radius does not have the relation to the hard substructure of the jet it is intended to.

More precisely, I-Pruning is defined as the subclass of pruned jets for which, during the sequential clustering, there was never a recombination with $\Delta R_{ij} > R_{prune}$ and $\min(p_{t,i}, p_{t,j}) > z_{prune} p_{t,(i+j)}$. The other situation, i.e. a pruned jet for which there was at least one recombination for which $\Delta R_{ij} > R_{prune}$ and $\min(p_{t,i}, p_{t,j}) > z_{prune} p_{t,(i+j)}$, corresponds to a genuine two-prong structure and is called Y-Pruning.

This distinction between I- and Y-Pruning is mostly irrelevant for boosted jet tagging. However, it has been shown to have an impact on the analytical behaviour of Pruning, with Y-Pruning being under better control and than I-Pruning, the latter adding an extra layer of complexity to the calculation. If one's goal is to reach some level of analytic control over groomed jets, Y-Pruning appears as a more natural choice than Pruning which also includes the contribution from I-Pruning.

Y-Splitter Y-Splitter is one of the very few tools proposed for boosted W-boson tagging at the LHC [4]. The idea is to recluster the constituents of the jet with the k_t algorithm and to undo the last step of the clustering. This gives two subjets j_1 and j_2. One then defines

$$y_{12} = \frac{k_{t,12}^2}{m_{12}^2} = \frac{\min(p_{t1}^2, p_{t2}^2)\Delta R_{12}^2}{m_{12}^2}, \tag{5.4}$$

similar to what has been used later in the MassDrop Tagger. One then imposes the cut $y > y_{cut}$ to require to hard prongs in the jet.[5] Note that similar quantities have been introduced as event shapes in e^+e^- collisions.

Johns Hopkins Top Tagger As its name suggests, this is a tagger meant to separate fat jets originating from the decay of boosted top quarks from the background made of light-quark jets. It was one of the first substructure techniques introduced in the context of LHC physics. The tagger aims at finding three hard prongs in the jet, corresponding to the $q\bar{q}b$ hard quarks produced by the hadronic decay of the top, adding constraints that two of the three prongs are compatible with a hadronically-decaying W boson. In practice, it proceeds as follows [127]:

1. If the initial jet has not been obtained by the Cambridge/Aachen algorithm, re-cluster the jet constituents using this algorithm,
2. *Primary decomposition*: as for the mMDT, we iteratively undo the last step of the Cambridge/Aachen clustering. The softer of the two subjets is discarded if its transverse momentum divided by the *initial* jet p_t is smaller than a parameter δ_p. The de-clustering procedure then continues with the harder subjet. This is repeated until one of our things happens: (i) both subjets are above δ_p, (ii) both subjets are below δ_p, (iii) the two subjets satisfy $|\Delta y| + |\Delta \phi| < \delta_r$, with δ_r another parameter of the tagger, or (iv) the subjet can no longer be declustered. In case (i) the two hard subjets are kept and further examined, in the other three cases, the jet is not tagged as a top candidate.
3. *Secondary decomposition*: with the two prongs found by the primary decomposition, repeat the declustering procedure as for the primary decomposition, still defining the δ_p condition with respect to the original jet p_t. This can result in either both prongs from the primary decomposition being declustered into two sub-prongs, only one prong being declustered, or none. When no further substructure is found in a primary prong, the primary prong is kept intact in the final list of prongs. When two sub-prongs are found both are kept in the final list

---

[5]A cut on y is roughly equivalent to a cut on the p_t fraction z. For example, for a jet made of two collinated partons carrying a momentum fraction z and $1 - z$ of the jet, one has $y = \frac{z}{1-z}$.

of prongs. Ultimately, this leads to two, three or four prongs emerging from the original jet. Only jets with three or four sub-prongs are then considered as top candidates, while the case with only two prongs is rejected.

4. *Kinematic cuts*: with the three or four prongs found from the secondary decomposition, impose additional kinematic conditions. First, the sum of the four-momenta of all the hard prongs should be close to the top mass. Then, there exists two prongs which invariant mass is close to the W mass. Finally we impose that the W helicity angle be consistent with a top decay. The W helicity angle, θ_h, is defined as the angle between the top direction and one of the W decay products, in the rest frame of the W. We impose $\cos(\theta_h) < 0.7$.[6]

The original paper suggested that the parameters should be adjusted according to the event's scalar E_T:

$$1 \text{ TeV} < E_T < 1.6 \text{ TeV} : R = 0.8, \quad \delta_p = 0.10, \quad \delta_r = 0.19, \tag{5.5}$$

$$1.6 \text{ TeV} < E_T < 2.6 \text{ TeV} : R = 0.6, \quad \delta_p = 0.05, \quad \delta_r = 0.19, \tag{5.6}$$

$$2.6 \text{ TeV} < E_T : R = 0.4, \quad \delta_p = 0.05, \quad \delta_r = 0.19. \tag{5.7}$$

The kinematic cuts are then adjusted based on the jet p_t:

$$p_t < 1 \text{ TeV}: 145 < m_{\text{top}} < 205 \text{ GeV}, \qquad 65 < m_W < 95 \text{ GeV}, \tag{5.8}$$

$$p_t > 1 \text{ TeV}: 145 < m_{\text{top}} < p_t/20 + 155 \text{ GeV}, \quad 65 < m_W < 70 + p_t/40 \text{ GeV}, \tag{5.9}$$

where m_{top} and m_W are the reconstructed top and W mass respectively.

The prong decomposition of the Johns Hopkins top tagger shared obvious similarities with the (modified) MassDrop Tagger introduced to tag Higgs bosons, in the sense that it follows the hardest branch on a Cambridge/Aachen clustering tree and imposes a hardness condition on the subjets. Since we now want to require three hard prongs in the jet, the de-clustering procedure is repeated twice. The main noticeable differences between the (modified) MassDrop Tagger and the Johns Hopkins top tagger is that the latter imposes a δ_p condition computed with respect to the original jet p_t while the mMDT imposes its z_{cut} condition computed as a fraction of the subjets parent's p_t. Note also the use of the Manhattan distance in the δ_r condition.

In practice, for a top efficiency between 20 and 40%, the Johns Hopkins top tagger achieves reductions of the background by a factor ~ 100 (remember these numbers should be squared for the efficiency to tag a $t\bar{t}$ pair).

---

[6]Top decays are almost isotropic and the helicity angle had an almost flat distribution, while for QCD jets, it diverges like $1/(1 - \cos(\theta_h))$.

CMS Top Tagger The CMS top tagger is essentially an adaptation of the Johns
Hopkins top tagger proposed by the CMS collaboration [128, 129]. Declustering
proceeds analogously to the Johns Hopkins top tagger—except for the two-prongs
distance condition which uses a p_t-dependent cut on the standard ΔR_{ij} subjet
distance—but the kinematic conditions are different. The detailed procedure works
as follows:

1. If needed, the initial jet is re-clustered using the Cambridge/Aachen algorithm.
2. *Primary decomposition*: the last step of the clustering is undone, giving two
   prongs. These two prongs are examined for the condition

$$p_t^{\text{prong}} > \delta_p \, p_t^{\text{jet}}, \tag{5.10}$$

where p_t^{jet} refers to the hard jet transverse momentum. δ_p is a parameter
which is usually taken as 0.05. If both prongs pass the cut then the "primary"
decomposition succeeds. If both prongs fail the cut then the jet is rejected i.e. is
not tagged as a top jet. If a single prong passes the cut the primary decomposition
recurses into the passed prong, until the decomposition succeeds or the whole jet
is rejected. Note that during the recurrence, p_t^{jet} (used in (5.10)) is kept as the
transverse momentum of the original jet.
3. *Secondary decomposition*: with the two prongs found by the primary decomposition, repeat the declustering procedure as for the primary decomposition, still
   defining the δ_p condition (5.10) with respect to the original jet p_t. This can result
   in either both prongs from the primary decomposition being declustered into
   two sub-prongs, only one prong being declustered, or none. When no further
   substructure is found in a primary prong, the primary prong is kept intact in the
   final list of prongs. When two sub-prongs are found both are kept in the final list
   of prongs. Ultimately, this leads to two, three or four prongs emerging from the
   original jet. Only jets with three or four sub-prongs are then considered as top
   candidates.
4. *Kinematic constraints*: taking the three highest p_t subjets (i.e. prongs) obtained
   by the declustering, find the minimum pairwise mass and require this to be related
   to the W mass, m_W, by imposing the condition $\min(m_{12}, m_{13}, m_{23}) > m_{\min}$ with
   $m_{\min} \lesssim m_W$. For practical applications, $m_{\min}$ is usually taken as 50 GeV.
5. Note that in the second version of the tagger [129], the decomposition procedure
   also imposes an angular cut: when examining the decomposition of a subjet S
   into two prongs i and j, the CMS tagger also requires $\Delta R_{ij} > 0.4 - A p_t^S$ where
   $\Delta R_{ij} = \sqrt{\Delta y_{ij}^2 + \Delta \phi_{ij}^2}$ and p_t^S refers to the transverse momentum of the subjet.
   The default value for A is $0.0004\,\text{GeV}^{-1}$. We note that without a ΔR condition in
   the decomposition of a cluster, the CMSTopTagger is collinear unsafe (see [130]
   for a discussion of this and proposed alternatives).

5.4 Radiation Constraints

The standard approach to constraining radiation inside a jet is to impose a cut on a *jet shape* which, similarly to event shapes in electron-positron collisions, is sensitive to the distribution of the particles in the jet (or in the event for the e^+e^- case). Over the past 10 years, several jet shapes have been introduced. In what follows, we review the most common ones.

5.4.1 Angularities and Generalised Angularities

The simplest family of jet shapes is probably the *generalised angularities* [131] defined as

$$\lambda_\beta^\kappa = \sum_{i \in \text{jet}} z_i^\kappa \left(\frac{\Delta R_{i,\text{jet}}}{R} \right)^\beta , \qquad (5.11)$$

where z_i is the jet transverse momentum fraction carried by the constituent i and $\Delta R_{i,\text{jet}}$ its distance to the jet axis:

$$z_i = \frac{p_{t,i}}{\sum_{j \in \text{jet}} p_{t,j}} \quad \text{and} \quad \Delta R_{i,\text{jet}}^2 = (y_i - y_{\text{jet}})^2 + (\Delta\phi - \phi_{\text{jet}})^2. \qquad (5.12)$$

Note that generalised angularities (and more in general, the other jet shapes presented later) can also be used for jets in e^+e^- collisions if we define $z_i = E_i/E_{\text{jet}}$ and replace $\Delta R_{i,\text{jet}}$ either by $\theta_{i,\text{jet}}$, the angle to the jet axis, or by $2\sin(\theta_{i,\text{jet}}/2) = \sqrt{2(1 - \cos\theta_{i,\text{jet}})}$.

Generalised angularities are collinear unsafe, except for the special case $\kappa = 1$ which corresponds to the IRC safe *angularities* [132, 133]:

$$\lambda_\beta \equiv \lambda_\beta^{(\kappa=1)}. \qquad (5.13)$$

The specific case $\beta = 1$ is sometimes referred to as *width* or *girth* or *broadening*, while $\beta = 2$ is closely related to the jet mass.[7]

Obviously, the more radiation there is in a jet, the larger generalised angularities are. Angularities and generalised angularities can therefore be seen as a measure of QCD radiation around the jet axis, i.e. as the radiation in a one-pronged jet. They are often used as a quark-gluon discriminator, where gluon-initiated jets would, on average, have larger angularity values that quark-initiated jets [134–136].

---

[7]It reduces to $\rho = m^2/(p_t R)^2$ in the limit of massless particles and small jet radius R.

For completeness, we note that the "jet" axis used to compute angularities can differ from the axis obtained via the initial jet clustering (usually the anti-k_t algorithm with jet radius R and E-scheme recombination). A typical example is the case of the jet width where using an axis defined with the E-scheme recombination introduces a sensitivity to recoil and complicates the analytic calculations of width. The workaround is to use a recoil-free axis, like the WTA recombination scheme. More generally, it is advisable to use the WTA axis for angular exponents $\beta \leq 1$. This is also valid for the other shapes defined below and we will adopt this choice when presenting analytic calculations.

There are at least two other examples of generalised angularities that, despite being IRC unsafe, are widely used in applications. The case $\beta = \kappa = 0$ corresponds to the *jet multiplicity*, and $\beta = 0, \kappa = 2$, which is related to p_t^D [137, 138]. Finally, generalised angularities can be defined as track-based observables by limiting the sum in Eq. (5.11) to the charged tracks (i.e. charged constituents) in the jet. Tracked-based angularities are advantageous in the context of pileup mitigation, because compared to neutral energy deposits in calorimeters, it is easier to separate tracks that originate from pileup vertices from tracks from the hard-interaction. The price we pay is that tracked-based observables are not IRC safe and theoretical predictions involve non-perturbative fragmentation functions [47, 48, 139].

5.4.2   N-subjettiness

As the name suggests, N-subjettiness [140] is a jet shape that aims to discriminate jets according to the number N of subjets they are made of. It takes inspiration from the event-shape N-jettiness [141]. In order to achieve this, a set of axes $a_1, \ldots, a_N$ is introduced (see below for a more precise definition) and the following jet shape is introduced[8]

$$\tau_N^{(\beta)} = \sum_{i \in \text{jet}} p_{ti} \min(\Delta R_{ia_1}^\beta, \ldots, \Delta R_{ia_N}^\beta), \qquad (5.14)$$

where β is a free parameter.[9] The axes a_i can be defined in several ways, the most common choices being the following:

- k_t **axes**: the jet is re-clustered with the k_t algorithm and the a_i are taken as the N exclusive jets.
- **WTA** k_t **axes**: the jet is re-clustered with the k_t algorithm, using the winner-take-all recombination scheme. The a_i are taken as the N exclusive jets. As for angularities, the use of the WTA axes guarantees a recoil-free observable.

---

[8]Equation (5.14) corresponds to the *un-normalised* definition of N-subjettiness. Alternatively, one can normalise τ_N by the jet scalar p_t, $\tilde{p}_t = \sum_{i \in \text{jet}} p_{ti}$, or, more simply, the jet p_t.

[9]Although it is strongly advised to specify the value of β one uses, $\beta = 1$ is often implicitly assumed in the literature.

- **generalised-k_t axes**: this is defined as above but now one uses the exclusive jets obtained with the generalised k_t algorithm. It is helpful to set the p parameter of the generalised k_t algorithm to $1/\beta$, so as to match the distance measure used for the clustering with the one used to compute τ_N. For $\beta < 1$ one would again use the WTA generalised-k_t axes.
- **minimal axes**: chose the axes a_i which minimise the value of τ_N. The minimum is found by iterating the minimisation procedure described in Ref. [142] starting with a set of seeds. It is often possible to find a less computer-expensive definition (amongst the other choices listed here) which would be as suitable to the minimal axes, both for phenomenological applications and for analytic calculations.
- **one-pass minimisation axes**: instead of running a full minimisation procedure as for the minimal axes, one can instead start from any other choice of axes listed above and run the minimisation procedure described in Ref. [142].

As for the angularities discussed in the previous section, τ_N is a measure of the radiation around the N axes $a_1, \ldots, a_N$. For a jet with N prongs, one expects $\tau_1, \ldots, \tau_{N-1}$ to be large and $\tau_{\geq N}$ to be small. The value of τ_N will also be larger when the prongs are gluons. For these reasons, the N-subjettiness ratio

$$\tau_{N,N-1}^{(\beta)} = \frac{\tau_N^{(\beta)}}{\tau_{N-1}^{(\beta)}} \tag{5.15}$$

is a good discriminating variable for N-prong signal jets against the QCD background. More precisely, one would impose a cut $\tau_{21}^{(\beta)} < \tau_{\text{cut}}$ to discriminate W/Z/H jets against QCD jets and $\tau_{32}^{(\beta)} < \tau_{\text{cut}}$ to discriminate top jets against QCD jets Although the most common use of N-subjettiness in the literature takes $\beta = 1$, there are also some motivations to use $\beta = 2$, see e.g. [121, 143].

5.4.3   Energy-Correlation Functions

Energy-correlation functions (ECFs) achieve essentially the same objective than N-subjettiness without requiring the selection of N reference axes. In their original formulation [143], they are defined as

$$e_2^{(\beta)} = \sum_{i<j\in\text{jet}} z_i z_j \, \Delta R_{ij}^\beta, \tag{5.16}$$

$$e_3^{(\beta)} = \sum_{i<j<k\in\text{jet}} z_i z_j z_k \, \Delta R_{ij}^\beta \Delta R_{jk}^\beta \Delta R_{ik}^\beta, \tag{5.17}$$

$$\vdots$$

$$e_N^{(\beta)} = \sum_{i_1<\ldots<i_N\in\text{jet}} \left(\prod_{j=1}^N z_{i_j} \right) \left(\prod_{k<\ell=1}^N \Delta R_{i_k i_\ell}^\beta \right), \tag{5.18}$$

with $z_i = p_{t,i} / \sum_j p_{t,j}$. Compared to N-subjettiness, energy-correlation functions have the advantage of not requiring a potentially delicate choice of reference axes. Furthermore, from an analytic viewpoint, they are insensitive to recoil for all values of the angular exponent β, allowing for an easier analytic treatment (although, as we have mentioned earlier, this issue can be alleviated in the N-subjettiness case by using WTA axes).

Generalised versions of the angularities have been introduced [144]. They still involve p_t weighted sums over pairs, triplets,... of particles but are built from other angular combinations:

$$_1e_2^{(\beta)} \equiv e_2, \tag{5.19}$$

$$_3e_3^{(\beta)} \equiv e_3, \tag{5.20}$$

$$_2e_3^{(\beta)} = \sum_{i<j<k\in\text{jet}} z_i z_j z_k \min\left(\Delta R_{ij}^\beta \Delta R_{ik}^\beta \Delta R_{ij}^\beta \Delta R_{jk}^\beta \Delta R_{ik}^\beta \Delta R_{jk}^\beta\right), \tag{5.21}$$

$$_1e_3^{(\beta)} = \sum_{i<j<k\in\text{jet}} z_i z_j z_k \min\left(\Delta R_{ij}^\beta, \Delta R_{ik}^\beta, \Delta R_{jk}^\beta\right), \tag{5.22}$$

$$\vdots$$

$$_ke_N^{(\beta)} = \sum_{i_1<...<i_N\in\text{jet}} \left(\prod_{j=1}^N z_{i_j}\right)\left(\prod_{\ell=1}^k \overset{\ell}{\underset{u<v\in\{i_1,...,i_N\}}{\min}} \Delta R_{uv}^\beta\right), \tag{5.23}$$

where $\overset{\ell}{\min}$ denotes the ℓ-th smallest number.

Similarly to N-subjettiness, in order to discriminate boosted massive particles from background QCD jets, we again introduce ratios of (generalised-)ECFs. Over the past few years, several combinations have been proposed. Examples of ratios of ECFs that are used as two-prong taggers include

$$C_2^{(\beta)} = \frac{_3e_3^{(\beta)}}{\left(_1e_2^{(\beta)}\right)^2} \equiv \frac{e_3^{(\beta)}}{\left(e_2^{(\beta)}\right)^2}, \qquad D_2^{(\beta)} = \frac{e_3^{(\beta)}}{\left(e_2^{(\beta)}\right)^3}, \tag{5.24}$$

$$N_2^{(\beta)} = \frac{_2e_3^{(\beta)}}{\left(e_2^{(\beta)}\right)^2}, \qquad M_2^{(\beta)} = \frac{_1e_3^{(\beta)}}{e_2^{(\beta)}},$$

while for three-prong tagging, one introduces [143–145]

$$C_3^{(\beta)} = \frac{e_4^{(\beta)} e_2^{(\beta)}}{\left(e_3^{(\beta)}\right)^2}, \qquad N_3 = \frac{_2e_4^{(\beta)}}{\left(_1e_3^{(\beta)}\right)^2}, \qquad M_3 = \frac{_1e_4^{(\beta)}}{_1e_3^{(\beta)}}, \tag{5.25}$$

$$D_3^{(\alpha,\beta,\gamma)} = \frac{e_4^{(\gamma)}\left(e_2^{(\alpha)}\right)^{\frac{3\gamma}{\alpha}}}{\left(e_3^{(\beta)}\right)^{\frac{3\gamma}{\beta}}} + \kappa_1\left(\frac{p_t^2}{m^2}\right)^{\frac{\alpha\gamma}{\beta}-\frac{\alpha}{2}}\frac{e_4^{(\gamma)}\left(e_2^{(\alpha)}\right)^{\frac{2\gamma}{\beta}-1}}{\left(e_3^{(\beta)}\right)^{\frac{2\gamma}{\beta}}}$$

$$+ \kappa_2\left(\frac{p_t^2}{m^2}\right)^{\frac{5\gamma}{2}-2\beta}\frac{e_4^{(\gamma)}\left(e_2^{(\alpha)}\right)^{\frac{2\beta}{\alpha}-\frac{\gamma}{\alpha}}}{\left(e_3^{(\beta)}\right)^2},$$

where κ_1 and κ_2 are $\mathcal{O}(1)$ constants.

In this series, the D family has typically a larger discriminating power, at the expense of being more sensitive to model-dependent soft contamination in the jet like the UE or pileup. Instead, the N family is closer to N-subjettiness, and the M family is less discriminating but more resilient against soft contamination in the jet.

Finally, we note that Energy Correlation functions have recently been extended into Energy Flow polynomials [146] which provide a linear basis for all infrared-and-collinear-safe jet substructure observables. These can then be used to design Energy Flow Networks [147] which are QCD-motivated machine-learning substructure tools.

5.4.4   Additional Shapes

Over the past decade, several other jet shapes have been introduced in the literature and studied by the LHC experiments. Since they tend to be less used than the ones introduced above, we just briefly list the most common ones below, without entering into a more detailed discussion.

Iterated SoftDrop This is related to Recursive SoftDrop introduced earlier. The idea is still to apply SoftDrop multiple times except that this time we will only follow the hardest branch in the recursion procedure [148]. This gives a list of branchings which pass the SoftDrop condition, $(z_1, \theta_1), \ldots, (z_n, \theta_n)$, from which we can build observables. The most interesting observable is probably the *Iterated SoftDrop multiplicity*, which is simply the number of branchings which have passed the SoftDrop condition and which is an efficient quark-gluon discriminator as we will show in Chap. 7. Alternatively, we can build Iterated SoftDrop angularities from the set of (z_i, θ_i). We note that for the Iterated SoftDrop multiplicity to be infrared and collinear safe, we need either to take a negative value of the SoftDrop parameter β or impose an explicit cut (in θ or in k_t).

Planar Flow Planar flow [133] (see also [149]) is defined as

$$Pf = \frac{4\det(I_\omega)}{\text{tr}^2(I_\omega)} = \frac{4\lambda_1\lambda_2}{(\lambda_1 + \lambda_2)^2} \quad \text{with} \quad I_\omega^{kl} = \sum_{i\in\text{jet}}\omega_i\frac{p_{i,k}}{\omega_i}\frac{p_{i,l}}{\omega_i}, \tag{5.26}$$

where m is the jet mass, ω_i is the energy of constituent i, $p_{i,k}$ the kth component of its transverse momentum with respect to the jet axis, and λ_1 and λ_2 are the eigenvalues of I_ω.

Planar flow is meant to tag object with 3-or-more-body decays. These would appear as a planar configuration with large values of Pf, while QCD jets tend to have a linear configuration and a small value of Pf. This is similar to the D-parameter in e^+e^- collisions. A boost-invariant version of planar flow can be defined as

$$ Pf_{\text{BI}} = \frac{4 \det(I_{\text{BI}})}{\text{tr}^2(I_{\text{BI}})} \qquad \text{with} \qquad I_{\text{BI}}^{\alpha\beta} = \sum_{i \in \text{jet}} p_{t,i}(\alpha_i - \alpha_{\text{jet}})(\beta_i - \beta_{\text{jet}}), \qquad (5.27) $$

where, now, α and β correspond either to the rapidity y or azimuth ϕ. We note that Pf and Pf_{BI} are quite sensitive to the UE and pileup activity in a jet (see e.g. [150]) making them difficult to use in experimental analyses. Since we will not come back to planar flow in our analytic calculations in the following chapters, let us mention that some fixed-order analytic results are available in the literature [151].

Q-jet Volatility The main idea behind Q-jet [152, 153] is to define jets as a set of multiple clustering trees (weighted by an appropriate metric) instead of a single one. A tree would be constructed using a modified pairwise-recombination algorithm working as follows:

1. for a set of particles at a given stage of the clustering, we first compute the k_t or Cambridge/Aachen set of distances d_{ij}. Let d_{min} be their minimum.
2. We then compute a set of weights w_{ij} for each pair and assign the probability $\Omega_{ij} = w_{ij}/\sum_{(ij)} w_{ij}$ to each pair. The weights are typically taken as

$$ w_{ij} = \exp\left(-\alpha \frac{d_{ij} - d_{\text{min}}}{d_{\text{min}}}\right) \qquad (5.28) $$

   where α is a parameter called rigidity.
3. we generate a random number used to select a pair (ij) with probability Ω_{ij}.
4. The pair is recombined and the procedure is iterated until no particles are left.

The algorithm is then repeated N_{tree} times. In the limit $\alpha \to \infty$ one recovers the standard clustering. In practice one usually takes $\alpha \simeq 0.01$ and $N_{\text{tree}} \gtrsim 50$ (typically 256).

Q-jets can then be used to compute jet physics observables, including substructure variables, by taking the statistical average over the many trees. New observables, related to the fact that we now have a distribution of trees, can also be considered. A powerful example is Q-jet volatility. It is defined by applying pruning together with Q-jet, i.e. imposing the pruning condition (see Sect. 5.3 above) on each of the clusterings trees, and then measuring the width of the resulting mass distribution:

$$V = \frac{\sqrt{\langle m^2 \rangle - \langle m \rangle^2}}{\langle m \rangle}. \tag{5.29}$$

When disentangling boosted W jets from background QCD jets, one would expect V to be smaller in W jets than in QCD jets, mostly because the former have a better-defined mass scale than the latter.

5.5 Combinations of Tools

A few methods commonly used in recent substructure works can be seen as combinations of ingredients borrowed from the two categories above. We list the most important ones in the next paragraphs.

Before doing so, we want to stress that substructure observables do not commute and therefore, when considering combinations tools, the order in which we apply the different algorithms does matter. For instance, when imposing both a condition on the "groomed" jet mass and on a jet shape, one would obtain different results if the jet shape is computed on the plain jet or on the groomed jet. A clear example of this is the combination of Y-splitter with trimming or the mMDT, where imposing the Y-splitter cut on the plain jet greatly improves performance. It is therefore important that the description of the tagging strategy clearly specify all the details of the combination including for example what jet, groomed or ungroomed, is used to compute jet shapes.

That said, while several specific combinations are worth mentioning, we limit ourselves to two-prong taggers:

ATLAS Two-Prong Tagger The standard algorithm adopted by ATLAS for Run-II of the LHC proceeds as follows. Trimming is applied to the jet, using the k_t algorithm with a trimming radius $R_{\text{trim}} = 0.2$ and an energy cut $f_{\text{trim}} = 0.05$. One then requires the trimmed mass to be between 65 and 105 GeV. One then computes $D_2^{(\beta=1)}$ on the trimmed jet and impose a cut on this variable.

CMS Two-Prong Taggers At LHC Run-II, CMS has used two different two-prong taggers. Both start by applying the mMDT to the anti-k_t ($R = 0.8$) jets with $z_{\text{cut}} = 0.1$ and require the mMDT mass to be between 65 and 105 GeV. At the beginning of Run-II, CMS was then computing the N-subjettiness $\tau_{21}^{(\beta=1,\text{plain})}$ ratio, using exclusive k_t axes to define the axes, on the plain jet, and imposing a $\tau_{21}^{(\beta=1,\text{plain})}$. More recently, they replaced the N-subjettiness cut by a cut on $N_2^{(\beta=1,\text{mMDT})}$ i.e. they impose instead a cut on an N_2 ratio computed of the groomed jet (see e.g. [154] for a recent analysis). In both cases, they used a decorrelated version of the shape (see below).

Decorrelated Taggers (DDT) Let us consider the combination of the mMDT with a cut on N-subjettiness. Because of the correlation between these two observables, a cut on the shape can significantly sculpt the jet mass distribution of the background, leading to a deterioration in performance. The idea behind the DDT procedure [155] is to instead substitute the cut on N-subjettiness, with a cut on a suitable combination of τ_{21} and of a function of the $\rho_{\mathrm{mMDT}} = m^2_{\mathrm{mMDT}}/(p_{t,\mathrm{mMDT}})^2$. This function is chosen such that the final background mass spectrum, after imposing a fixed cut on the decorrelated shape, is flat. The flatness of the background makes it easier for searches where the mass of the signal is unknown (or when the p_t of the jet can widely vary). In Ref. [155], it was shown that $\tau_{21} - \mathrm{cst.} \times \log(\rho_{\mathrm{mMDT}})$, with the constant determined from the ρ_{mMDT} dependence of the average τ_{21} value was giving good results. This can easily be extended to other combinations. For example, CMS has recently used a decorrelated N_2 variable defined as $N_2^{\mathrm{DDT}} = N_2 - N_2(\text{cut at } 5\%)$ where $N_2(\text{cut at } 5\%)$ corresponds to the value of a cut on N_2 that would give a 5% background rate. We also refer to [156, 157] for examples where decorrelated shapes are built analytically.

Dichroic Ratios There is a conceptual difference between imposing the shape cut on the plain jet or on the groomed jet. Since shapes measure the soft radiation at large angles, one should expect a better performance when the cut is imposed on the plain jet, since any grooming algorithm would have, by definition, eliminated some of the soft-and-large-angle radiation. Conversely, this very same soft-and-large-angle part of the phase-space is the one which is most sensitive to the UE and pileup, so computing the shape on the groomed jet would be more resilient to these effects. Recently, it was proposed to adopt a hybrid, *dichroic*, approach. The starting point is the observation that the shapes are meant to constrain additional radiation, on top of the two hard prongs. For ratios the sensitivity to the extra radiation is usually captured by the numerator, e.g. τ_2, while the denominator (e.g. τ_1) is mostly sensitive to the two hard prongs.

That said, the first step of a full two-prong tagger is usually to apply a groomer/prong-finder, say the mMDT, in order to resolve the two-prong structure of the jet and impose a cut on the mass. On then imposes a radiation constrain. For the latter it is therefore natural to compute the denominator of the shape, here τ_1, (sensitive to the two hard prongs) on the result of the groomer/prong-finder jet. In order to retain information about the soft-and-large-angle radiation in the jet (where one expects discriminating power), one then wishes to compute the numerator of the shape, here τ_2, on a larger jet. The latter can be either the plain jet or, if we want a compromise between performance and soft resilience, a lightly-groomed jet like a SoftDrop jet with a positive β (typically $\beta = 2$) and a smallish z_{cut}. This defines the dichroic N-subjettiness ratio [121]

$$\tau_{21}^{(\beta=2,\mathrm{dichroic})} = \frac{\tau_2^{(\beta=2,\mathrm{loose\ grooming})}}{\tau_1^{(\beta=2,\mathrm{tight\ grooming})}}, \tag{5.30}$$

which has been shown to give good results on Monte-Carlo simulation and analytic calculations. Although it was initially introduced for $\beta = 2$ N-subjettiness, it can be applied to other shapes as well.

Additional Remarks Besides the specific prescriptions discussed above, it is helpful to keep a few generic ideas in mind when combining different substructure tools:

- When the M, N and U series of generalised angularities have been introduced, their combination with a grooming procedure was also discussed. We therefore encourage the reader interested in additional details to refer to Ref. [144].
- In a similar spirit, combining a Y-splitter cut, computed on the plain jet, with a grooming technique, such as trimming or the mMDT, for the measurement of the jet mass has been shown [120, 158] to provide nice improvements both over Y-splitter alone—owing to a reduced sensitivity to soft non-perturbative effects—and over grooming alone—owing to a larger suppression of the QCD background.
- When one uses tagging techniques based on radiation constraints, one may want to first run a SoftDrop grooming procedure with positive β, i.e. as a groomer, so as to limit the sensitivity to pileup and the Underlying Event, while keeping some of the soft-and-large-angle radiation for the radiation constraint.[10]

Finally, we note that a systematic and extensive investigation of the tagging performance and resilience to non-perturbative effects obtained when combining one of many prong finders with one of many radiation constraints has been investigated in the context of the Les Houches *Physics at TeV colliders* workshop in 2017. We will briefly cover that study in Chap. 8, but we refer to Section III.2 of [122] for more details (cf. also our discussion on performance assessment in Sect. 5.2).

5.6 Other Important Tools

As all classifications, separation of substructure tools in prong finders and radiations constraints has its limits and some methods do not obviously fall in either category. In this section we list the most important ones.

---

[10]Overall, it appears natural to use in parallel negative, or zero, β as a tool to identify the two-prong structure and positive β with a jet shape, to impose a cut on radiation.

5.6.1 Shower Deconstruction

Given a set of four-momenta p_N of the N measured final state objects, one can associate probabilities $P(p_N|S)$ and $P(p_N|B)$ that it was initiated by a signal (S) or background (B) process respectively. From these probabilities one can build an ideal classifier[11]

$$\chi(p_N) = \frac{P(p_N|S)}{P(p_N|B)}. \tag{5.31}$$

This fundamental observation is also the foundation of the so-called matrix-element method [159, 160], used in various applications in particle phenomenology with fixed-order matrix elements [161–163].

Shower deconstruction also relies on Eq. (5.31) to separate boosted jets from signal from boosted background jets. As discussed in Chap. 4, the probabilities $P(p_N|S)$ and $P(p_N|B)$ cannot reliably be computed at fixed order due to the disparate scales in the process. Instead one makes use of all-order calculations in QCD to compute $\chi(p_N)$.

In practice, shower deconstruction considers all possible splittings of the set $\{p_N\} = \{p_I\} \cup \{p_F\}$ into initial an final-state radiation. For each such splitting it then considers all possible shower histories, taking into account all possible parton-flavor assignments, that could lead to the final state $\{p_N\}$. A weight can then be calculated in perturbative QCD (see below) for each history and the probabilities $P(p_N|S, B)$ are taken as the sum of all the weights associated with $\{p_N\}$ under the signal or background hypothesis. To compute the weight for a given history, one uses a Feynman-diagrammatic approach [164, 165] where each vertex receives a factor of the form $H e^{-R}$ with H a partonic splitting probability at a given virtuality and e^{-R} is a Sudakov factor, built from the splitting probability H which accounts for the fact that the splitting did not happen at a larger virtuality. The specific form of H depends on the splitting at hand, using e.g. Eq. (2.15) and Eqs. (2.27)–(2.28) for QCD branchings, however retaining full mass dependency for the partons involved, thereby reaching a modified leading-logarithmic accuracy and the full LO matrix element for the decay of W/Z/H bosons or top quarks.

At the moment, probabilities are available for massive or massless quark, gluons, hadronically-decaying electroweak W/Z/H bosons and hadronically-decaying top quarks. This makes shower deconstruction readily available for quark-gluon discrimination, W/Z/H boosted bosons tagging and top tagging.

Note also that including all the constituents of the jet can quickly become prohibitive due to the large number of possible histories. A workaround is to first recluster the jet into small subjets and use those subjets as an input to shower deconstruction.

---

[11]The Neyman-Pearson Lemma proves formally that χ, as defined in Eq. (5.31), is an ideal classifier.

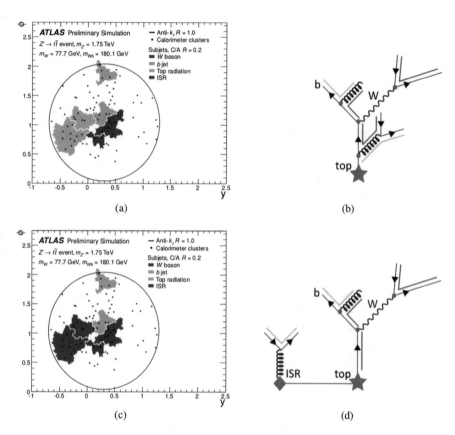

Fig. 5.2 The figure (taken from Ref. [166]) illustrates how shower deconstruction works as a top tagger. The left-hand panel shows the energy depositions in the rapidity-azimuth plane, while the left-hand panel shows the corresponding most-likely shower histories. The coloured lines in the right panels indicate which partons are colour-connected in the respective shower histories

To illustrate the process, Fig. 5.2 (taken from Ref. [166]) shows the two histories out of more than 1500 with the largest probabilities for a particular simulated $Z' \rightarrow t\bar{t}$ event, where the leading large-radius jet (anti-k_t, with $R = 1$) in this event was reclustered into six subjets (using the Cambridge/Aachen algorithm $R = 0.2$). The left plots show energy deposits, while the right panels show the actual histories.

5.6.2   HEP Top Tagger

The HEP top tagger was first designed to reconstruct mildly boosted top quarks in a busy event environment, i.e. for the reconstruction of top quarks in the process $pp \rightarrow \bar{t}th$ with semi-leptonic top quark decays and H $\rightarrow \bar{b}b$ [167]. The hadronically top was expected to be boosted in the p_t range around 250–500 GeV.

This first incarnation of the tagger was augmented by cuts on observables that were manifestly Lorentz-invariant, and thus boosting between reference frames were no longer necessary. It proceeds as follows (see Appendix A of [168]):

1. one first defines the fat jets with the Cambridge/Aachen algorithm with $R = 1.5$,
2. for a given fat jet, one recursively undoes the last step of the clustering, i.e. decluster the jet j into subjets j_1 and j_2 with $m_{j_1} > m_{j_2}$, until we observe a mass-drop $m_{j_1} < 0.8 m_j$. When the mass-drop condition is not met, one carries on with the declustering procedure with j_1.
3. For subjets which have passed the mass-drop condition and which satisfy $m_j > 30$ GeV, one further decomposes the subjet recursively into smaller subjets.
4. The next step is to apply a filter similarly to what is done by the Mass-Drop Tagger. One considers all pairs of hard subjets, defining a filtering radius $R_{\text{filt}} = \min(0.3, \Delta R_{ij})$. We then add a third hard subjet—considering again all possible combinations—and apply the filter on the three hard subjets keeping (at most) the 5 hardest pieces and use that to compute the jet mass. Amongst all possible triplets of the original hard subjets, we keep the combination for which the jet mass—calculated after filtering—gives the mass closest to the top mass and is within a mass window around the true top mass, e.g. in the range 150–200 GeV.
5. Out of the 5 filtered pieces, one extracts a subset of 3 pieces, j_1, j_2, j_3, ordered in p_t and accept it as a top candidate if the masses satisfy at least one of the following 3 criteria:

$$0.2 < \arctan\left(\frac{m_{13}}{m_{12}}\right) < 1.3 \quad \text{and} \quad R_{\min} < \frac{m_{23}}{m_{123}} < R_{\max} \tag{5.32}$$

$$R_{\min}^2\left(1 + \frac{m_{13}^2}{m_{123}^2}\right) < 1 - \frac{m_{23}^2}{m_{123}^2} < R_{\max}^2\left(1 + \frac{m_{13}^2}{m_{123}^2}\right) \quad \text{and} \quad \frac{m_{23}}{m_{123}} > 0.35$$

$$R_{\min}^2\left(1 + \frac{m_{12}^2}{m_{123}^2}\right) < 1 - \frac{m_{23}^2}{m_{123}^2} < R_{\max}^2\left(1 + \frac{m_{12}^2}{m_{123}^2}\right) \quad \text{and} \quad \frac{m_{23}}{m_{123}} > 0.35,$$

with $R_{\min} = 0.85 \, m_W/m_t$ and $R_{\max} = 1.15 \, m_W/m_t$.
6. the combined p_t of the 3 subjets constructed in the previous step is imposed to be at least 200 GeV.

Physically, the first three steps above try to decompose a massive object into its hard partons, in a spirit similar to what the mass-drop condition used in the MassDrop tagger does. The filtering step also plays the same role of further cleaning the contamination from the Underlying Event as in the MassDrop tagger. Finally, the set of constraints in (5.32) is meant as a cut on the 3-subjets, mimicking a 3-parton system, to match the kinematics of a top decay and further suppress the QCD background. The whole procedure can be visualised as shown in Fig. 5.3.

Version 2 of the HEPTopTagger [169] brings several improvements by using an extended set of variables and cuts. We just list those modifications without entering into the details. First, it introduces a variable radius by repeatedly reducing the jet radius, starting from $R = 1.5$, until we see a drop in the reconstructed top mass.

HEP Top Tagger details

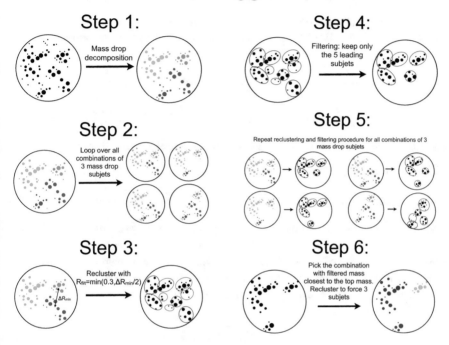

Fig. 5.3 Visualisation of the HEP top tagger algorithm

This is meant to reduce possible combinatorial effects where the softest of the W decays is mistaken with a hardish QCD subjet in the fat top candidate jet. Then, the tagger includes additional shape variables:

- N-subjettiness values for $\beta = 1$ computed both on the plain, ungroomed, jet and on the filtered jet
- Q-jet information: the reconstructed top mass obtained from 100 Q-jet histories based on the Cambridge/Aachen algorithm with $\alpha = 1$, as well as the fraction of positive top tags one would obtain with version 1 of the HEPTopTagger.

In the end, the tagger uses a multivariate (Boosted Decision Tree) analysis based on the series of kinematic variables—subjet transverse momenta and masses—the optimal jet radius, and the shape values.

5.6.3 The Lund Jet Plane

In Sect. 4.2, we have introduced the Lund plane as a graphical representation convenient for resummation calculations. It has actually been realised recently that, in the context of jet substructure, it was possible to promote this idea to a genuine observable [65].

In practice, one reclusters the constituents of the jet with the Cambridge/Aachen algorithm and apply the following iterative procedure, starting with the full jet:

1. decluster the jet in two subjets p_i and p_j, with $p_{ti} > p_{tj}$.
2. with the idea that this corresponds to the emission of p_j from an emitter $p_i + p_j$, one defines the following variables:

$$\Delta \equiv \Delta R_{ij}, \qquad k_t \equiv p_{tj}\Delta, \qquad m^2 \equiv (p_i + p_j)^2 \qquad (5.33)$$

$$z \equiv \frac{p_{tj}}{p_{ti} + p_{tj}}, \qquad \kappa \equiv z\Delta, \qquad \psi \equiv \tan^{-1}\frac{y_j - y_i}{\phi_j - \phi_i} \qquad (5.34)$$

3. Iterate the procedure by going back to step one for the harder subjet p_i.

This construction produce an ordered list of tuples

$$\mathcal{L}_{\text{primary}} \equiv \left[\mathcal{T}^{(1)}, \ldots, \mathcal{T}^{(n)}\right] \qquad \text{with } \mathcal{T}^{(i)} \equiv \left\{\Delta^{(i)}, k_k^{(i)}, \ldots\right\}, \qquad (5.35)$$

where $\mathcal{T}^{(i)}$ corresponds to the ith step of the declustering procedure. In particular, the set of pairs $(\log(1/\Delta^{(i)}), \log(k_t^{(i)}))$ corresponds to a representation of all the primary emissions of a given jet in the Lund-plane representation of Sect. 4.2 (cf. Fig. 4.1). This provides a overview of the internal structure of the jet.

Figure 5.4 shows the average Lund plane density

$$\rho(\Delta, k_t) \equiv \frac{1}{N_{\text{jet}}} \frac{dn_{\text{emissions}}}{d\log(1/\Delta)d\log(k_t)} \qquad (5.36)$$

Fig. 5.4 The average primary Lund plane density, ρ, for jets clustered with the C/A algorithm with $R = 1$. We selected jets having $p_t > 2$ TeV and $|y| < 2.5$

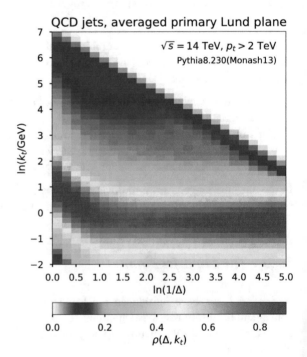

obtained from Pythia simulations using a dijet event sample. The main regions labelled in Fig. 4.1 can be clearly identified in Fig. 5.4.

As shown in Ref. [65], the Lund jet plane can be used for measurements and for constraining Monte-Carlo generators. Furthermore, boosted object taggers can be built from the Lund plane either via a log-likelihood approach, or using machine learning techniques.

5.7   Code Availability

An essential component of a successful jet substructure algorithm, is its availability. Therefore, for completeness, we list below where one can find the implementation of the tools presented above.

Tool	Code
Mass-Drop Tagger	`MassDropTagger` class in `FastJet`
modified Mass-Drop Tagger	`ModifiedMassDropTagger` class in the `RecursiveTools` `FastJet` contrib
SoftDrop	`SoftDrop` class in the `RecursiveTools` `FastJet` contrib
Recursive SoftDrop	`RecursiveSoftDrop` class in the `RecursiveTools` `FastJet` contrib
Filtering	`Filter` class in `FastJet` (use `SelectorNHardest`)
Trimming	`Filter` class in `FastJet` (use `SelectorPtFractionMin`)
Pruning	`Pruner` class in `FastJet`
I and Y-Pruning	Not available per se but can be implemented as a derived class of `Pruner`
Johns Hopkins top tagger	`JHTopTagger` class in `FastJet`
CMS top tagger	As part of CMS-SW (see Ref. [170])
Generalised angularities	No know public standard implementation
N-subjettiness	`Nsubjettiness` `FastJet` contrib
Energy Correlation Functions	`EnergyCorrelator` `FastJet` contrib
HEPTopTagger	Code available from Ref. [171]
Shower Deconstruction	Code available from Ref. [172]

Let us conclude this chapter with a more general remark. Grooming techniques might at first sight be similar to pileup mitigation techniques. They however target a different goal: while pileup mitigation techniques aim at correcting for the average effect of pileup, grooming techniques reduce the overall sensitivity to pileup. In practice, this means that, unless one first applies an event-wide pileup mitigation technique such as SoftKiller [173] or PUPPI [174], grooming techniques should in principle be supplemented by pileup subtraction, like the area–median [64, 150, 175, 176]. Many tools provide hooks to combine them with pileup subtraction.

Chapter 6
Calculations for the Jet Mass
with Grooming

In this chapter we will revisit the calculations performed in Chap. 4 and extend them in order to describe jet mass distributions with grooming algorithms. In what follows, we are not going to present state-of-the art theoretical calculations, but instead we aim to keep the our discussion as simple as possible. Therefore, the theoretical accuracy of the calculations that we will present will be the minimum one which is required to capture the essential feature of the distributions. We will mostly concentrate of QCD jets, which present the most interesting and intricate features, while a discussion about jets originated to a boosted heavy particles will be presented in Sect. 6.4.

6.1 mMDT/SoftDrop Mass

The first calculation we perform is that of the invariant mass distribution of a jet after the mMDT/SoftDrop algorithm has been applied. As we have already mentioned, the SoftDrop drop algorithm reduces to mMDT when the angular exponent β is set to zero. Therefore, in order to keep our notation light we are going to generically refer to the algorithm as SoftDrop (SD) and it is understood that the $\beta = 0$ case corresponds to mMDT.

In the next subsection, we do the calculation at leading order in the strong coupling constant. This simple example will allow us to see the large logarithms that appear and we will turn to their resummation in the next subsection.

© Springer Nature Switzerland AG 2019
S. Marzani et al., *Looking Inside Jets*, Lecture Notes in Physics 958,
https://doi.org/10.1007/978-3-030-15709-8_6

Fig. 6.1 Diagram
contributing to the
leading-order mass
distribution

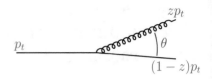

6.1.1 LO Calculation

At zeroth order in α_s, the jet mass is always zero. To obtain a non-trivial mass,
we therefore need to consider a high-energy parton, say a quark for definiteness,
radiating an extra gluon, as depicted in Fig. 6.1. We want to focus on the boosted jet
limit and highlight large logarithms of m/p_t, with p_t the transverse momentum
of the initial quark, which arise in the perturbative series expansion. At the
leading-logarithmic accuracy we are interested in, we can work in the collinear
approximation where the gluon emission angle θ is small.[1] The gluon is set to carry
a fraction z of the quark momentum, leaving a fraction $1 - z$ for the recoiling quark
after the emission.

When applying SD, the jet is split into two subjets, one with the quark and one
with the gluon which is tested for the SD condition. Two situations can occur: (1)
either the splitting passes the SD condition, i.e. $z > z_{\text{cut}}(\theta/R)^\beta$ in which case the
quark-gluon system is retained by the SD procedure and the (squared) jet mass is
given by

$$m^2 = z(1 - z)\theta^2 p_t^2, \tag{6.1}$$

or (2) the condition is failed in which case only the harder of the quark and the gluon
is kept and the jet mass vanishes. The mass distribution at LO is therefore given by

$$\frac{m^2}{\sigma} \frac{d\sigma^{(\text{LO})}}{dm^2} = \frac{\alpha_s}{2\pi} \int_0^{R^2} \frac{d\theta^2}{\theta^2} \int_0^1 dz \, P_q(z) m^2 \delta(m^2 - z(1-z)\theta^2 p_t^2)$$

$$\Theta(z > z_{\text{cut}}(\theta/R)^\beta), \tag{6.2}$$

where $P_q(z)$ is the quark splitting function.

The mass constraint can be used to perform the integration over θ, and the
constraint $\theta < R$ means we have to impose $z(1 - z) > \rho$ where we have introduced
the dimensionless variable

$$\rho = \frac{m^2}{p_t^2 R^2}. \tag{6.3}$$

---

[1] Alternatively, we can assume a small jet radius R so that corrections beyond the collinear
approximation are suppressed by powers of R. Note also that in the case of the mMDT jet mass,
the SD condition actually gets rid of this contribution so that the collinear approximation remains
valid at higher logarithmic accuracy.

Up to power corrections in ρ, i.e. in the groomed jet mass, we can neglect the factor $1 - z$ in this constraint. We are therefore left with

$$\frac{\rho}{\sigma} \frac{d\sigma^{(\text{LO})}}{d\rho} = \frac{\alpha_s}{2\pi} \int_\rho^1 dz \, P_q(z) \Theta\big(z^{2+\beta/2} > z_{\text{cut}}^{2/(2+\beta)} \rho^{\beta/(2+\beta)}\big). \tag{6.4}$$

In the remaining integration over z, the SD constraint is only relevant for $\rho < z_{\text{cut}}$ and we get

$$\frac{\rho}{\sigma} \frac{d\sigma_{\text{SD}}^{(\text{LO})}}{d\rho} = \begin{cases} \frac{\alpha_s C_F}{\pi} \left[\log\left(\frac{1}{\rho}\right) - \frac{3}{4} \right], & \text{if } \rho > z_{\text{cut}}, \\ \frac{\alpha_s C_F}{\pi} \left[\frac{\beta}{2+\beta} \log\left(\frac{1}{\rho}\right) + \frac{2}{2+\beta} \log\left(\frac{1}{z_{\text{cut}}}\right) - \frac{3}{4} \right], & \text{if } \rho < z_{\text{cut}}, \end{cases}$$
$$\tag{6.5}$$

again up to power corrections in ρ.[2] The above result exhibits two different regimes: when the jet mass is not very small $\rho > z_{\text{cut}}$, SD is inactive and one recovers the plain, i.e. ungroomed jet-mass distribution discussed in Chap. 4. However, when the mass becomes smaller, $\rho < z_{\text{cut}}$, SD becomes active, as manifested here under the form of a larger cut on the z integration in Eq. (6.4).

As mentioned in Chap. 4, it is usual to work with the cumulative distribution. At $\mathcal{O}(\alpha_s)$, we find

$$\Sigma_{\text{SD}}^{(\text{LO})}(\rho) = \frac{1}{\sigma_0} \int_0^\rho d\rho' \frac{d\sigma}{d\rho'} = 1 - \frac{1}{\sigma_0} \int_\rho^1 d\rho' \frac{d\sigma}{d\rho'} \tag{6.6}$$

$$= \begin{cases} 1 - \frac{\alpha_s C_F}{\pi} \left[\frac{1}{2} \log^2\left(\frac{1}{\rho}\right) - \frac{3}{4} \log\left(\frac{1}{\rho}\right) \right], & \text{if } \rho > z_{\text{cut}}, \\ 1 - \frac{\alpha_s C_F}{\pi} \left[\frac{1}{2} \log^2\left(\frac{1}{\rho}\right) - \frac{1}{2+\beta} \log^2\left(\frac{z_{\text{cut}}}{\rho}\right) - \frac{3}{4} \log\left(\frac{1}{\rho}\right) \right], & \text{if } \rho < z_{\text{cut}}. \end{cases}$$

In going from the first to the second equality, one could either argue that the probability is conserved (i.e. the mass is either larger or smaller than ρ), or realise that $\frac{d\sigma}{d\rho'}$ also has a virtual contribution at $\rho' = 0$ which, up to subleading power corrections, can be written as

$$\left. \frac{d\sigma^{(\text{LO})}}{d\rho'} \right|_{\text{virt.}} = - \left(\int_0^1 d\rho \frac{d\sigma}{d\rho} \right) \delta(\rho').$$

More importantly, the results above clearly show that a gluon emission comes with large logarithms of the jet mass on top the expected power of α_s. When the jet mass becomes sufficiently small, this is no longer a small quantity and one needs

---

[2]Technically, for mMDT, this result is valid up to power corrections in z_{cut}. These corrections can be included and resummed [123, 177] but we will assume small z_{cut} here and neglect them.

to resum gluon emissions to all orders. We do that in the next section. There are however a few interesting points we can already highlight now. For example, we see that the dominant logarithms in $\Sigma(\rho)$ are double logarithms of the jet mass. These are associated with the emission of a gluon which is both soft and collinear. The subleading single-logarithmic contribution comes here from a hard and collinear gluon emission. Then, one expects the SD condition to be less effective as β increases. This is indeed what one sees here since one tends to the plain jet mass distribution in the limit $\beta \to \infty$. Conversely, for $\beta = 0$, the double logarithm of the jet mass disappears—going back to Eq. (6.4) the z integration is cut at z_{cut} for $\beta = 0$, meaning that the soft emissions only produce a logarithm of z_{cut} instead of a combination of $\log(z_{\text{cut}})$ and $\log(\rho)$ for the generic case—leaving a single-logarithmic dominant term, which is purely collinear.

We conclude this section with a discussion about soft emissions at large angles. These have not been included in the calculation above where we have worked in the collinear, small R, approximation. However, as seen in Chap. 4 (see e.g. Eq. (4.7)), soft emissions at finite angles can also give single-logarithmic contributions. This will no longer be the case in the region where SD is active. To see this, imagine that we have a soft emission passing the SD condition and dominating the jet mass. This implies $\rho = z(\theta/R)^2$ and $z > z_{\text{cut}}(\theta/R)^\beta$, from which one easily deduces $\theta < R(\rho/z_{\text{cut}})^{1/(2+\beta)}$. A contribution at a finite angle (i.e. not enhanced by a collinear $d\theta/\theta$) would therefore be suppressed by a power of ρ. Similarly, one can show that non-global logarithms are also suppressed by SD. This is a fundamental analytic property of SD, namely that it suppresses soft-and-large-angle gluon emissions so that observables can (usually) be computed in the collinear limit. We will come back to that point in the next section.

6.1.2   Resummation of the mMDT/SoftDrop Mass Distribution

We now move to the all-order resummation of the logarithms of the SD jet mass distribution. We target a modified leading-logarithmic accuracy, i.e. include the leading double-logarithmic terms as well as the hard-collinear single-logarithmic contributions.

In an all-order calculation, one has two types of contributions to consider. First, real emissions which fail the SD condition will be groomed away by the SD procedure[3] and will therefore not contribute to the jet mass. They will therefore cancel explicitly against the corresponding virtual corrections. We are therefore left with the case of the real gluons which pass the SD condition and the associated

---

[3]Strictly speaking, since SD stops the first time the condition is passed, this is only true for gluons at angles larger than the first emission passing the SD condition. However, such gluons cannot dominate the jet mass and so can be neglected. It is worth noting that for more complicated quantities, like jet shapes computed on a SD jet, this effect would have to be taken into account.

virtual emissions. These gluons will contribute to the jet mass. The situation here is therefore exactly as the one discussed in Sect. 4.2 for the case of the plain jet mass but now restricted to the gluons passing the SD condition.

At the end of the day, this means that, if we want to compute the cumulative distribution $\Sigma_{SD}(\rho)$, we have to veto all real emissions that, while passing the SD condition, would give a "mass" larger than ρ. Real emissions outside the SD region and emissions at smaller mass do not contribute to the jet mass[4] and cancel against virtual corrections. We are therefore left with a "standard" Sudakov-type factor

$$\Sigma_{SD}(\rho) = \exp\left[-R_{SD}(\rho)\right], \tag{6.7}$$

with (measuring the angles in units of the jet radius R for convenience and $i = q, g$)

$$R_{SD}(\rho) = \int_0^1 \frac{d\theta^2}{\theta^2} \, dz \, P_i(z) \frac{\alpha_s(z\theta p_t R)}{2\pi} \Theta(z\theta^2 > \rho)\Theta(z > z_{cut}\theta^\beta). \tag{6.8}$$

In a fixed-coupling approximation, R_{SD} is the same as the one-gluon emission result, Eq. (6.6). Including running-coupling corrections is straightforward. We choose the hard scale to be $p_t R$ and we write

$$\alpha_s(z\theta p_t R) = \frac{\alpha_s(p_t R)}{1 + 2\alpha_s\beta_0 \log(z\theta)}, \tag{6.9}$$

and we perform the integration keeping only the leading double-logarithmic contributions from soft-and-collinear emissions as well as hard-collinear branchings. For $\rho < z_{cut}$, we obtain

$$R_{SD}^{(LL)}(\rho) = \frac{C_i}{2\pi\alpha_s\beta_0^2}\left[\frac{2+\beta}{1+\beta}W\left(1 - \frac{\lambda_c + (1+\beta)\lambda_\rho}{2+\beta}\right) - \frac{W(1-\lambda_c)}{1+\beta}\right.$$
$$\left. - 2W\left(1 - \frac{\lambda_\rho}{2}\right) - 2\alpha_s\beta_0 B_i \log\left(1 - \frac{\lambda_\rho}{2}\right)\right], \tag{6.10}$$

with

$$\lambda_\rho = 2\alpha_s\beta_0 \log(1/\rho), \qquad \lambda_c = 2\alpha_s\beta_0 \log(1/z_{cut}), \qquad \text{and} \quad W(x) = x\log(x).$$

The first line in Eq. (6.10) corresponds to the double logarithms, while the second line comes from hard-collinear splittings. This expression covers both the case of quark- and gluon-initiated jets, with the only difference between the two are the

---

[4]At full single-logarithmic accuracy, one would also get a contribution with multiple emissions contributing to the jet mass, These emission would again have to pass the SD condition and their resummation goes exactly as for the plain jet, yielding a factor $\exp(-R')/\Gamma(1 + R')$ with R' the derivative of the SD radiator given below.

overall colour factor ($C_i = C_F$ for quarks and $C_i = C_A$ for gluons) and the contribution from hard-collinear splittings ($B_i = B_q$ or $B_i = B_g$, see Appendix A). As before, we recover the plain-jet case in the limit $\beta \to \infty$, while the distribution becomes single-logarithmic for the mMDT case, i.e. $\beta = 0$. Note that it might be convenient to reabsorb the contribution from hard-collinear splittings, the last term of Eq. (6.10), directly into the double-logarithmic contribution. This gives an expression equivalent to Eq. (6.10) up to NNLL corrections:

$$R_{\text{SD}}^{(\text{LL}')}(\rho) = \frac{C_i}{2\pi\alpha_s\beta_0^2}\left[\frac{2+\beta}{1+\beta}W\left(1 - \frac{\lambda_c + (1+\beta)\lambda_\rho}{2+\beta}\right) - \frac{W(1-\lambda_c)}{1+\beta}\right.$$
$$\left. -2W\left(1 - \frac{\lambda_\rho + \lambda_B}{2}\right) + W(1-\lambda_B)\right], \qquad (6.11)$$

with $\lambda_B = -2\alpha_s\beta_0 B_i$. The pros and cons of this alternative treatment of the B term are further discussed in Appendix A. More generally, the B terms can systematically be inserted in the LL contributions by replacing the $z < 1$ kinematic boundary by $z < \exp(B_i)$. This is the approach we have adopted for all the plots obtained from analytic calculations in this chapter.

The above results can easily be represented using Lund diagrams (cf. Sect. 4.2). This is done in Fig. 6.2. compared to the plain jet mass, only the emissions above the SD condition have to be vetoed. This corresponds to the shaded red region on the plot, therefore corresponding to the radiator R_{SD}. Similarly, its derivative with respect to $\log(1/\rho)$, R'_{SD}, is the weight associated with having an emission passing the SD condition and satisfying $z\theta^2 = \rho$, and is represented by the solid red

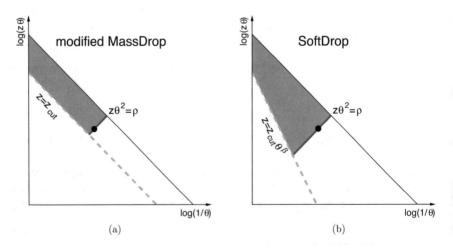

(a)   (b)

Fig. 6.2 Lund diagrams for the groomed jet mass distribution at LL for mMDT (**a**) and generic SD (**b**). The solid green line represents the edge of the SD region, corresponding to the condition $z = z_{\text{cut}}\theta^\beta$. The solid red line corresponds to emissions yielding the requested jet mass, i.e. satisfying $z\theta^2 = \rho$. The shaded red area is the vetoed area associated with the Sudakov suppression

line in Fig. 6.2. From both the analytic results and the simple Lund diagrams, one clearly sees that the smaller β, the more aggressively one grooms soft-and-large-angle emissions. Furthermore, when β decreases, both R_{SD} and R'_{SD} decrease.

6.2 Other Examples: Trimming and Pruning

Amongst the taggers and groomers introduced in Chap. 5, the modified Mass-Drop Tagger and Soft Drop are the ones with the simpler analytic structure. It is however possible to obtain results for other groomers/taggers as well. In this section we give a brief overview of the mass distribution one would obtain after applying trimming or pruning, as initially calculated in Ref. [123]. We refer to Sect. 5.3 for a description of the substructure tools.

6.2.1 Trimming

Leading-Order Result As above, we start with $\mathcal{O}(\alpha_s)$ calculation. Therefore, we consider a single soft and collinear gluon emission in the jet, emitted from a high-energy quark at an angle θ and carrying a fraction z of the leading parton's momentum. For the jet mass to be non-zero, the emission needs to be kept in the trimmed jet. If the emission is clustered in the same subjet as the leading parton, it will automatically be kept; otherwise, if it is in its own subjet, it will only be kept if it carries a fraction of the total jet p_t larger than f_{trim}. After adding together real and virtual contribution. the LO contribution to the cumulative distribution is[5]:

$$\Sigma_{\text{trim}}^{(LO)}(\rho) = 1 - \frac{\alpha_s}{2\pi} \int_0^1 \frac{d\theta^2}{\theta^2} \int_0^1 dz \, P_q(z) \, \Theta(z\theta^2 > \rho) \, \Theta(z > f_{\text{trim}} \text{ or } \theta < r_{\text{trim}}),$$

(6.12)

where we have introduced $r_{\text{trim}} = R_{\text{trim}}/R$. We note that the above expression differs from the mMDT/SD case only by the tagger/groomer condition. Therefore, if we are only interested in terms enhanced by logarithms of ρ, f_{trim} or r_{trim}, we can easily follow the same approach as in Sect. 6.1.1 and get

$$\Sigma_{\text{trim}}^{(LO)}(\rho) = 1 - \frac{\alpha_s C_F}{\pi} \left[\frac{1}{2} \log^2 \left(\frac{1}{\rho} \right) - \frac{1}{2} \log^2 \left(\frac{f_{\text{trim}}}{\rho} \right) \Theta(\rho < f_{\text{trim}}) \right. \tag{6.13}$$

$$\left. + \frac{1}{2} \log^2 \left(\frac{f_{\text{trim}} r_{\text{trim}}^2}{\rho} \right) \Theta(\rho < f_{\text{trim}} r_{\text{trim}}^2) - \frac{3}{4} \log \left(\frac{1}{\rho} \right) \right],$$

---

[5]For brevity, the notation $\Theta(a \text{ or } b)$ is one if either a or b is satisfied and 0 is none of a and b are satisfied. It can be rewritten as $\Theta(a \text{ or } b) = \Theta(a) + (1 - \Theta(a))\Theta(b) = \Theta(b) + (1 - \Theta(b))\Theta(a)$.

This results is very similar to what was obtained for the mMDT, i.e. Eq. (6.6) with $\beta = 0$, with one striking difference: there is an additional transition point at $\rho = f_{\text{trim}} r_{\text{trim}}^2$. For $f_{\text{trim}} r_{\text{trim}}^2 < \rho < f_{\text{trim}}$, the distribution is single-logarithmic and is the same as what one gets for $\rho < z_{\text{cut}}$ in the mMDT case with the replacement $z_{\text{cut}} \rightarrow f_{\text{trim}}$. However, at lower ρ, one has an extra contribution, $\frac{1}{2} \log^2(f_{\text{trim}} r_{\text{trim}}^2 / \rho)$, corresponding to a typical plain-jet double-logarithmic contribution (albeit for a jet of smaller radius).

For completeness, we also give the results for the differential mass distribution at leading order, which reads

$$\frac{\rho}{\sigma} \frac{d\sigma_{\text{trim}}^{(\text{LO})}}{d\rho} = \begin{cases} \frac{\alpha_s C_F}{\pi} \left[\log\left(\frac{1}{\rho}\right) - \frac{3}{4} \right] & \text{if } \rho \geq f_{\text{trim}}, \\ \frac{\alpha_s C_F}{\pi} \left[\log\left(\frac{1}{f_{\text{trim}}}\right) - \frac{3}{4} \right] & \text{if } f_{\text{trim}} r_{\text{trim}}^2 \leq \rho < f_{\text{trim}} \\ \frac{\alpha_s C_F}{\pi} \left[\log\left(\frac{r_{\text{trim}}^2}{\rho}\right) - \frac{3}{4} \right] & \text{if } \rho < f_{\text{trim}} r_{\text{trim}}^2. \end{cases} \qquad (6.14)$$

All-Order Resummation As for the SoftDrop case, it is relatively easy to show that the all-order resummed result is simply the exponential of the one-gluon emission result (including running-coupling corrections which we shall not explicitly calculate here). We therefore get

$$\Sigma_{\text{trim}}^{(\text{LL})}(\rho) = \exp\left[-R_{\text{trim}}(\rho) \right], \qquad (6.15)$$

with, up to running-coupling corrections,

$$R_{\text{trim}}(\rho) = 1 - \Sigma_{\text{trim}}^{(\text{LO})}(\rho).$$

It is also informative to look at the corresponding Lund diagram, plotted in Fig. 6.3. Compared to Fig. 6.2a, we explicitly see the emergence of a transition point at $\rho = f_{\text{trim}} R_{\text{trim}}^2$ and a double-logarithmic behaviour in ρ at smaller masses. This is associated with the trimming radius R_{trim} and the fact that emissions at angles smaller than R_{trim} will be kept in the groomed jet regardless of their momentum fraction. This was different in the mMDT case where these emissions would still be subject to the mMDT z_{cut} constraint.

Finally, we can argue that this extra transition point is pathological and a strong motivation to prefer the mMDT and SoftDrop over trimming. Indeed, this transition point produces a kink in the mass spectrum (see also Sect. 6.3 below), smeared by subleading contributions. Finding a possible signal in this region, or using this mass domain as a side-band for a signal in an adjacent mass window, would then become much more complex, if not impossible. Additionally, this region would also receive single-logarithmic contributions from soft-and-large angle emissions and non-global logarithms (albeit suppressed by R_{trim}^2) which were absent in the SoftDrop case.

Fig. 6.3 Lund diagrams for the trimmed jet mass distribution at LL for mMDT (left) and generic SD (right). The solid green and blue lines represents the edge of the trimming region, respectively representing the $z = f_{trim}$ and $\theta = R_{trim}$ conditions. The solid red line corresponds to emissions yielding the requested jet mass, i.e. satisfying $z\theta^2 = \rho$. The shaded red area is the vetoed area associated with the Sudakov suppression

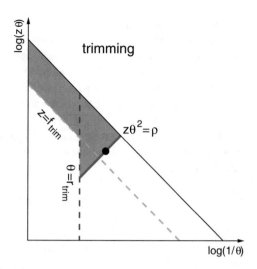

Thus, all these factors render the calculation of the trimmed mass spectra of the same degree of complexity as the plain jet mass, if not worse because of the presence of the transition points. On the other hand, the analytic structure we have found for SoftDrop was remarkable simpler and therefore amenable for precision calculations.

6.2.2 Pruning

In this section, we show explicitly that the case of pruning is more complex but can be simplified by introducing instead the Y-pruning variant. Since the main issue of pruning does not appear in a LO calculation, we will briefly discuss its origin at NLO, without providing an explicit calculation. For simplicity, we take $f_{prune} = \frac{1}{2}$, so that the pruning radius is given by $R_{prune} = m_{jet}/p_{t,jet}$ and we introduce $r_{prune} = R_{prune}/R$.

Leading-Order Result For a single soft-and-collinear emission of momentum fraction z and emission angle θ, the jet mass is given by $z\theta^2$, meaning that the pruning radius will be set to $R_{prune} = \sqrt{z}\theta$ which is always smaller than θ. The emission will therefore be kept in the pruned jet only if $z > z_{prune}$. This give exactly the same result as for mMDT, with z_{cut} replaced by z_{prune}:

$$\Sigma_{prune}^{(LO)}(\rho) = \Sigma_{mMDT}^{(LO)}(\rho)\Big|_{z_{cut} \to z_{prune}}, \qquad (6.16)$$

where we recall that mMDT corresponds to SoftDrop with angular exponent $\beta = 0$.

Behaviour at Higher Orders The pruning behaviour becomes significantly more complicated beyond LO. Let us give an explicit example. At NLO, we should consider situations where we have two real emissions, 1 and 2, with respective momentum fractions z_1 and z_2 and emission angles θ_1 and θ_2 with respect to the leading parton (one should as well include the cases with one or two virtual emissions). Without loss of generality, we can assume that $z_1\theta_1^2 \gg z_2\theta_2^2$, with the strong ordering sufficient to capture the leading logarithms of the jet mass we are interested in. Emission 1 therefore dominates the (plain) jet mass and sets the pruning radius to $R_{\text{prune}} = \sqrt{z_1}\theta_1$. The complication comes from the fact that emission 1 itself may be groomed away by pruning, i.e. have $z_1 < z_{\text{prune}}$, in which case, the jet mass will only be non-zero if emission 2 is kept by pruning and this is ensured by the condition

$$\Theta(z_2 > z_{\text{prune}} \text{ or } \theta_2^2 < z_1\theta_1^2),$$

which depends on z_1. As we will see below, this is not a show-stopper to resum the pruned jet mass distribution to all orders but we definitely depart from the simple Sudakov exponentiation seen for SoftDrop, Eq. (6.7), and trimming, Eq. (6.15).

All-Order Resummation To construct the all-order result, it is easier to consider the differential jet mass distribution. Let us then denote by "in" the emission that dominates the pruned jet, carrying a fraction z_{in} of the jet p_t and emitted at an angle θ_{in}, such that $\rho = z_{\text{in}}\theta_{\text{in}}^2$.

The pruning radius in units of the original jet radius is given by $r_{\text{prune}}^2 = R_{\text{prune}}^2/R^2 = m_{\text{jet}}^2/(p_{t,\text{jet}}R)^2$ which is set by the emission dominating the plain jet mass. We thus need to consider two cases: (1) there are no emissions in the plain jet with $z\theta^2 > z_{\text{in}}\theta_{\text{in}}^2$, (2) there is at least an emission in the plain jet with $z\theta^2 > z_{\text{in}}\theta_{\text{in}}^2$, and we call emission "out" the one with the largest $z\theta^2$, introducing $\rho_{\text{out}} = z_{\text{out}}\theta_{\text{out}}^2$. The corresponding Lund diagram is shown in Fig. 6.4a. In the first case, the pruning radius is set by emission one, $r_{\text{prune}}^2 = \rho < \theta_{\text{in}}^2$. To be in the pruned jet, the "in" emission should therefore satisfy $z_{\text{in}} > z_{\text{prune}}$. We get an associated Sudakov suppression $\exp(-R_{\text{plain}}(\rho))$ since we must veto emissions at larger mass than ρ both in the pruned jet and in the plain jet. In the second case, the pruning radius is set by the "out" emission, i.e. $r_{\text{prune}}^2 = \rho_{\text{out}} > \rho$. For $\rho > z_{\text{prune}}\rho_{\text{out}}$, the pruning condition is then $z_{\text{in}} > z_{\text{prune}}$ (shown in Fig. 6.4b), while for $\rho < z_{\text{prune}}\rho_{\text{out}}$ it becomes $z_{\text{in}} > r_{\text{prune}} = \rho_{\text{out}}$ (Fig. 6.4c). The Sudakov receives two different contributions: one from inside the pruning region, down to the scale ρ, represented by the red shaded are in Fig. 6.4, and one from outside the pruning region, the grey area in Fig. 6.4. Note that since $\rho_{\text{out}} < z_{\text{prune}}$, the situation $\rho < z_{\text{prune}}\rho_{\text{out}}$ only happens for $\rho < z_{\text{prune}}^2$, yielding a transition point at $\rho = z_{\text{prune}}^2$.

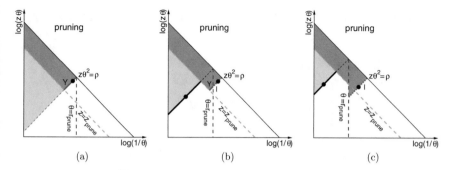

Fig. 6.4 Lund diagrams for the groomed jet mass distribution at LL with pruning in three different kinematic configurations. In case (**a**), the emission that dominates the plain jet mass (and hence sets the pruning radius) also has $z > z_{\text{prune}}$. In cases (**b**) and (**c**), the emission that dominates the plain mass and sets the pruning radius has $z < z_{\text{prune}}$ and does not pass the pruning condition. Another emission at lower mass dominates the pruned jet mass. This emission can either be constrained by the condition $z > z_{\text{prune}}$, case (**b**), or by the condition $\theta > r_{\text{prune}}$, case (**c**). For each of the three cases, we indicate the contributions to Y- and I-pruning

For $z_{\text{prune}}^2 < \rho < z_{\text{prune}}$, the sum over the two regions can be written as

$$
\frac{\rho}{\sigma} \frac{d\sigma_{\text{prune}}}{d\rho} = \int_{z_{\text{prune}}}^{1} dz_{\text{in}} P_i(z_{\text{in}}) \frac{\alpha_s}{2\pi} e^{-R_{\text{in}}(\rho)}
$$

$$
\left[e^{-R_{\text{out}}(\rho)} + \int_{\rho}^{z_{\text{prune}}} \frac{d\rho_{\text{out}}}{\rho_{\text{out}}} R'_{\text{out}}(\rho_{\text{out}}) e^{-R_{\text{out}}(\rho_{\text{out}})} \right]
$$

$$
= R'_{\text{in}}(\rho) e^{-R_{\text{in}}(\rho)}, \quad \text{with } \rho > z_{\text{prune}}^2, \tag{6.17}
$$

where we have introduced the radiators

$$
R_{\text{in}}(\rho) = R_{\text{mMDT}}(\rho), \tag{6.18}
$$

$$
R_{\text{out}}(\rho) = R_{\text{plain}}(\rho) - R_{\text{mMDT}}(\rho), \tag{6.19}
$$

where R_{mMDT} is obtained by setting $\beta = 0$ in Eq. (6.10). The radiators correspond respectively to the region kept (the shaded red area of Fig. 6.4) and rejected (the grey area of Fig. 6.4) by pruning. As long as the pruning condition is only $z > z_{\text{prune}}$, as in the above case, the R_{in} Sudakov is the same as the mMDT Sudakov, Eqs. (6.8) and (6.10). The R'_{out} factor in the first line of Eq. (6.17) corresponds to the integral over the momentum fraction of the emission outside the pruning region, represented by the solid black line in Fig. 6.4. After integration, we find that the pruned jet mass distribution is identical to the mMDT mass distribution for $\rho > z_{\text{prune}}^2$.

The situation for $\rho < z_{\text{prune}}^2$ is more involved as one now has to include the situation from Fig. 6.4c as well. In that case, the R_{in} Sudakov gets an additional contribution and the lower bound of the z_{in} integration extends down to r_{prune}. We then write

$$
\frac{\rho \, d\sigma_{\text{prune}}}{\sigma \quad d\rho} = \int_{z_{\text{prune}}}^{1} dz_{\text{in}} \, P_i(z_{\text{in}}) \frac{\alpha_s}{2\pi} e^{-R_{\text{in}}(\rho)}
$$

$$
\left[e^{-R_{\text{out}}(\rho)} + \int_{\rho}^{\rho/z_{\text{prune}}} \frac{d\rho_{\text{out}}}{\rho_{\text{out}}} R'_{\text{out}}(\rho_{\text{out}}) e^{-R_{\text{out}}(\rho_{\text{out}})} \right]
$$

$$
+ \int_{\rho/z_{\text{prune}}}^{z_{\text{prune}}} \frac{d\rho_{\text{out}}}{\rho_{\text{out}}} R'_{\text{out}} e^{-R_{\text{out}}(\rho)-R_{\text{in}}(\rho;\rho_{\text{out}})} \int_{\rho_{\text{out}}}^{1} dz_{\text{in}} \, P_i(z_{\text{in}}) \frac{\alpha_s}{2\pi}
$$

$$
= R'_{\text{in}}(\rho) e^{-R_{\text{in}}(\rho)-R_{\text{out}}(\frac{\rho}{z_{\text{prune}}})}
$$

$$
+ \int_{\rho/z_{\text{prune}}}^{z_{\text{prune}}} \frac{d\rho_{\text{out}}}{\rho_{\text{out}}} R'_{\text{out}} e^{-R_{\text{out}}(\rho)-R_{\text{in}}(\rho;\rho_{\text{out}})} \int_{\rho_{\text{out}}}^{1} dz_{\text{in}} \, P_i(z_{\text{in}}) \frac{\alpha_s}{2\pi}
$$

$$
\tag{6.20}
$$

with the new radiator

$$
R_{\text{in}}(\rho; \rho_{\text{out}}) = \int_0^1 \frac{d\theta^2}{\theta^2} dz \, P_i(z) \frac{\alpha_s}{2\pi} \Theta(z\theta^2 > \rho) \Theta(z > \min(z_{\text{prune}}, \rho_{\text{out}})),
$$

$$
\tag{6.21}
$$

corresponding to the shaded red region of Fig. 6.4c. Some simplifications and approximations can be done at fixed coupling but the main message here is that at small ρ, $\rho < z_{\text{prune}}^2$, the pruned mass distribution no longer involves a Sudakov which is the simple exponentiation of the one-gluon-emission result. In that region, one is left with an additional integration over the plain jet mass ρ_{out} which gives a Sudakov with double logarithms of the pruned jet mass ρ.

Y-pruning and I-pruning The main complication of pruning originates from the situation depicted in Fig. 6.4c, where the pruning radius is set by an emission which is groomed away. and the pruned mass is dominated by an emission at an angle smaller than the pruning radius. In this situation the prune radius is anomalously large because it is not set by hard splitting, as one would physically expect from pruning, especially when it is used as a two-prong tagger and the pruned jet is characterised by just one hard prong, hence the name I-pruning. Conversely, Y-pruning configurations are characterised by a hard $1 \to 2$ splitting. It is therefore interesting to compute the jet mass for Y-pruning.

The situation of Fig. 6.4c where the emission that dominates the pruned jet mass always has $\theta < r_{\text{prune}}$ is of the I-pruning type and does not contribute at all to Y-pruning. This is already a great simplification since, for example, all the expressions will now involve the simple $R_{\text{in}}(\rho)$ Sudakov and no longer $R_{\text{in}}(\rho'; \rho_{\text{out}})$. Furthermore, for the cases where the emission setting the pruned mass also sets the plain jet mass, Fig. 6.4a, we always have $z_{\text{in}} > z_{\text{prune}}$ and $\theta_{\text{in}} > r_{\text{prune}} = \sqrt{z_{\text{in}}} \theta_{\text{in}}$, meaning that this situation is always of the Y-pruning type.

Unfortunately, there is a price to pay for the remaining contribution, Fig. 6.4b for which, as indicated on the figure, one only gets a jet contributing to Y-pruning for smaller values of z_{in}, namely for $z_{\text{prune}} < z_{\text{in}} < \rho/\rho_{\text{out}}$.[6] Taking this into account, the Y-pruned jet mass distribution can be written as (assuming $\rho < z_{\text{prune}}$)

$$\frac{\rho}{\sigma}\frac{d\sigma}{d\rho} = \int_{z_{\text{prune}}}^{1} dz_{\text{in}} P_i(z_{\text{in}}) \frac{\alpha_s}{2\pi} e^{-R_{\text{plain}}(\rho)}$$

$$+ \int_{\rho}^{\min(z_{\text{prune}}, \rho/z_{\text{prune}})} \frac{d\rho_{\text{out}}}{\rho_{\text{out}}} R'_{\text{out}}(\rho_{\text{out}}) \qquad (6.22)$$

$$e^{-R_{\text{out}}(\rho_{\text{out}}) - R_{\text{in}}(\rho)} \int_{z_{\text{prune}}}^{\rho/\rho_{\text{out}}} dz_{\text{in}} P_i(z_{\text{in}}).$$

Inverting the two integrations on the second line, one can perform explicitly the integration over ρ_{out} and keep only an integration over z_{in}:

$$\frac{\rho}{\sigma}\frac{d\sigma}{d\rho} = \int_{z_{\text{prune}}}^{1} dz_{\text{in}} P_i(z_{\text{in}}) \frac{\alpha_s}{2\pi} e^{-R_{\text{in}}(\rho) - R_{\text{out}}(\min(z_{\text{prune}}, \rho/z_{\text{in}}))}. \qquad (6.23)$$

The Sudakov in the z_{in} integrand has a few interesting properties. First, for $\rho/z_{\text{in}} > z_{\text{prune}}$, it involves $R_{\text{out}}(z_{\text{prune}}) = 0$ and we recover a behaviour similar to what was seen for the mMDT. For $\rho/z_{\text{in}} < z_{\text{prune}}$, which is always the case for $\rho < z_{\text{prune}}^2$, R_{out} then becomes double-logarithmic in ρ.

6.2.3 Non-perturbative Corrections in Groomed Distributions

In Sect. 4.2.4 we have provided a rough estimate of the value of the jet mass at which the distribution becomes sensitive to non-perturbative physics. It is instructive to study revisit that calculation and see how this non-perturbative transition point changes if grooming techniques are applied.

We start by considering trimming. Assuming that we are in the $\rho < f_{\text{trim}} r_{\text{trim}}^2$ region, the situation is analogous to the plain jet mass and the mass m at which one becomes sensitive to non-perturbative effects is the same as Eq. (4.42) but with the jet radius substituted by the trimming radius

$$m^2 \simeq \frac{\mu_{\text{NP}}}{p_t R_{\text{trim}}} p_t^2 R_{\text{trim}}^2 = \mu_{\text{NP}} p_t R_{\text{trim}}, \qquad (6.24)$$

---

[6]This argument is not entirely true since even for $z_{\text{in}} > \rho/\rho_{\text{out}}$ we could still have another emission with $z > z_{\text{prune}}$, $\theta > r_{\text{prune}}$ and $z\theta^2 < \rho$. Such a contribution would only give terms proportional to $\alpha_s \log^2(z_{\text{prune}})$ i.e. not enhanced by any logarithm of the jet mass. We therefore neglect these contributions here.

where, compared to Eq. (4.42), we have switched to hadron-collider variables and used p_t rather than E_J. For pruning (both Y- and I-configuration), the non-perturbative transition point is formally the same as the plain jet mass, essentially because it is the latter that sets the pruning radius. Note however, that the size of non-perturbative corrections can differ with respect to the plain mass and one does expect pruning to achieve a significant reduction.

For the SoftDrop case we need a new calculation. We have to work out when an emission of constant $\rho = z\theta^2$, and passing the SoftDrop condition, first crosses into the non-perturbative region $z\theta < \tilde{\mu} = \frac{\mu_{NP}}{p_t R}$. This happens at the maximum allowed (rescaled) angle ($\theta = 1$ for the plain mass) which is determined by the SoftDrop condition $z = z_{cut}\theta^\beta$. We obtain $\rho \simeq \tilde{\mu} \left(\frac{\tilde{\mu}}{z_{cut}}\right)^{\frac{1}{1+\beta}}$, which implies

$$m^2 \simeq \mu_{NP}^{\frac{2+\beta}{1+\beta}} z_{cut}^{\frac{-1}{1+\beta}} (p_t R)^{\frac{\beta}{1+\beta}}. \tag{6.25}$$

Compared to the plain jet mass case, Eq. (4.42), the (squared) mass at which one becomes sensitive to non-perturbative effects is therefore smaller by a factor $\left(\frac{\mu_{NP}}{z_{cut} p_t R}\right)^{\frac{1}{1+\beta}}$. Once again, we note that the mMDT limit $\beta = 0$ is particularly intriguing as the p_t dependence disappears from Eq. (6.25).

6.2.4 Summary and Generic Overview

To conclude this section on analytic calculations, we summarise the basic analytic properties of the groomers/taggers in Table 6.1.

Table 6.1 Summary of the basic analytic properties of taggers

Groomer/ tagger	Transition points	Exponentiates	Largest logs	Soft logs	Non-global logs	Non-pert m^2 scale
Plain	–	Yes	$\alpha_s^n L^{2n}$	Yes	Yes	$\mu_{NP} p_t R$
mMDT	z_{cut}	Yes	$\alpha_s^n L^n$	No	No	μ_{NP}^2 / z_{cut}
SoftDrop	z_{cut}	Yes	$\alpha_s^n L^{2n}$	No	No	$\left(\frac{\mu_{NP}^{2+\beta}(p_t R)^\beta}{z_{cut}}\right)^{\frac{1}{1+\beta}}$
Trimming	$f_{trim}, f_{trim} r_{trim}^2$	Yes	$\alpha_s^n L^{2n}$	Yes	Yes	$\mu_{NP} p_t R_{trim}$
Pruning	z_{prune}, z_{prune}^2	No	$\alpha_s^n L^{2n}$	Yes	Yes	$\mu_{NP} p_t R$
Y-pruning	z_{prune}	No	$\alpha_s^n L^{2n-1}$	Yes	Yes	$\mu_{NP} p_t R$

Here, $L = \log(\rho)$. By soft logs we mean logarithmic contributions originating from soft emissions at finite angle. We note that SoftDrop does retain soft/collinear contributions (hence the double logarithmic behaviour), while mMDT only keeps hard-collinear radiation

A few key observations can be made.

- The modified MassDropTagger and SoftDrop groom soft radiations at all angular scales, i.e. without stopping at a given subjet radius. This has the consequence that they are insensitive to soft gluon emissions at finite angles and have no non-global logarithms.
- Another consequence of the absence of a subjet radius for mMDT and SD is that they are free of transition points beyond the one at $\rho = z_{\text{cut}}$. This is also the case of Y-pruning. Transition points can have subtle consequences in phenomenological applications and are therefore best avoided if possible. Furthermore, as we shall see explicitly in comparisons to Monte Carlo simulation in Sect. 6.3.2 below, for heavily boosted bosons these transition points can be around the electroweak scale and therefore have delicate side-effects when used in tagging boosted electroweak bosons.
- the simple symmetry cut of the mMDT, independent on the emission angles, translates into a perturbative logarithmic series where there are no double logarithms and the leading contributions are single logarithms of the jet mass. Although this translates into a smaller Sudakov suppression of the QCD background, this has the advantage of being theoretically simple. This, together with the fact that the mMDT strongly reduces the sensitivity to non-perturbative effects (see Sect. 6.3 below), is why it is a tool with a great potential for precision physics at the LHC.
- For many of the groomers we have studied, the resummed result has a simple structure where the one-gluon-emission expression simply exponentiates. The main exception to that is pruning which does not exponentiate. The situation is partially alleviated in the Y-pruning case.

6.3 Comparison to Monte Carlo

Now that we have obtained resummed results it is instructive to compare our findings to Monte Carlo simulations, which are ubiquitously used in phenomenology. We first do that at leading order to explicitly test the appearance of logarithms of the jet mass and check our control over the associated coefficients. We then move to a comparison to parton-shower simulations. In this case we will also discuss the impact of non-perturbative effects.

6.3.1 Comparisons at Leading Order

An simple test of the above substructure calculations is to verify that they do reproduce the logarithmic behaviour of a fixed-order calculation. To this purpose, we can use the Event2 [35, 36] generator. Although the program generates e^+e^-

Fig. 6.5 Comparison of the (normalised) mass distribution obtained at leading order, $\mathcal{O}(\alpha_s)$, between Event2 (solid lines) and our analytic expectations (dashed lines). The distribution is shown for both the plain jet (red) and a series of groomers: SoftDrop with $\beta = 2$ (green), mMDT (blue) and trimming with $R_{\text{trim}} = 0.2$ (black). The lower panel shows the difference between Event2 and the associated analytic expectation

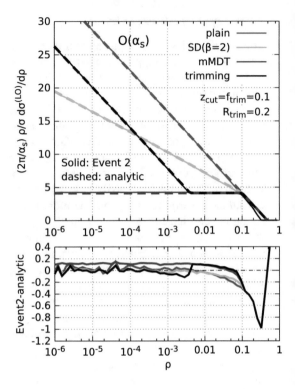

collisions, one can simulate quark jets of a given p_t (at $y = \pi = 0$) by rotating the whole event so that the thrust axis (or, alternatively the axis of the reference $q\bar{q}$ event generated by Event2) aligns with the x axis. We then cluster the jets with the anti-k_t algorithm [63][7] with $R = 1$ (cf. Chap. 3). We then apply any groomer to the resulting jets and measure the groomed jet mass. In practice, we have used mMDT with $z_{\text{cut}} = 0.1$, SoftDrop with $\beta = 2$ and $z_{\text{cut}} = 0.1$ and trimming with $f_{\text{trim}} = 0.1$ and $R_{\text{trim}} = 0.2$. In this section, we focus on the lowest non-trivial order of perturbation theory, $\mathcal{O}(\alpha_s)$. since we need at least 2 partons in a jet if we want a non-zero mass, it is sufficient to consider the real gluon emissions, i.e. $e^+ e^- \to q\bar{q}g$ events.

Figure 6.5 shows the mass distribution for a few selected groomers, together with the analytic calculations from above, expanded at order α_s. For SoftDrop, this is given by Eq. (6.5), while for trimming by Eq. (6.14). At $\mathcal{O}(\alpha_s)$, pruning and Y-pruning coincide with the mMDT and are therefore not showed. The bottom panel of the plot shows the difference between the Event2 simulations and the analytic results.

All the features discussed in this chapter are clearly visible on this plot: the transition points, at $\rho = z_{\text{cut}}$ for SD and at $\rho = f_{\text{trim}}$ and $\rho = f_{\text{trim}} r_{\text{trim}}^2$ for

---

[7]At leading order, $\mathcal{O}(\alpha_s)$, one could equivalently use any algorithm in the generalised-k_t family.

trimming, are present in the exact Event2 simulation; the effect of grooming is clearly visible at small ρ, with a reduction of the cross-section; the reduced $\log(\rho)$ contribution with SoftDrop and the absence of the $\log(\rho)$ enhancement for mMDT; the equivalence of trimming and mMDT in the intermediate ρ region; and the reappearance of the plain-mass-like $\log(\rho)$ contribution at small ρ for trimming.

Comparing the asymptotic behaviour at small mass to our analytic calculation, we first see that the leading logarithmic behaviour, i.e. the $\log(\rho)$ contribution, is correctly reproduced. This is visible on the bottom panel of Fig. 6.5 where all curves tend to a constant at small ρ. Furthermore, for trimming and SoftDrop, the analytic calculation also captures the constant term—$B_q = -\frac{3}{4}$ coming from hard-collinear branchings—and the difference between Event2 and the analytic results vanishes at small ρ. Although it is a bit delicate to see it on the figure, in the case of the plain, ungroomed, jet, this difference is only going to a non-zero constant at small ρ, because our calculation is missing a finite R^2 contribution coming from the emission of a soft gluon at a large angle. Finally, in the case of the mMDT, this difference is clearly different from 0 at small ρ. This originates from the fact that our analytic calculation in Sect. 6.1 has assumed $z_{\text{cut}} \ll 1$. For a finite value of z_{cut}, one has to keep the full z dependence in the splitting function which, at $\mathcal{O}(\alpha_s)$ means

$$
\frac{\rho}{\sigma}\frac{d\sigma_{\text{mMDT}}}{d\rho} = \frac{\alpha_s C_F}{2\pi} \int_{z_{\text{cut}}}^{1-z_{\text{cut}}} dz \frac{1+(1-z)^2}{z}
$$
$$
= \frac{\alpha_s C_F}{\pi} \left[\log\left(\frac{1-z_{\text{cut}}}{z_{\text{cut}}}\right) - \frac{3}{4}(1-2z_{\text{cut}}) \right]. \tag{6.26}
$$

Finite z_{cut} effects are of then given by $\frac{\alpha_s C_F}{\pi}\left[\frac{3}{2}z_{\text{cut}} - \log(1-z_{\text{cut}})\right]$. Pulling out an $\frac{\alpha_s}{2\pi}$ factor as done in Event2 and in Fig. 6.5, this gives a difference around 0.12 for our choice of $z_{\text{cut}} = 0.1$, which corresponds to what is observed on the plot.

6.3.2 Comparisons with Parton Shower

Setup We now compare our all-order results, including running coupling, to a full parton-shower simulation. For this, we use the Pythia8 [178] generator, in its Monash13 tune [179] at parton level. We generate dijet events at $\sqrt{s} = 13$ TeV, restricting the hard matrix element to $qq \rightarrow qq$ processes. Jets are reconstructed with the anti-k_t algorithm, as implemented in FastJet [56, 77], with $R = 1$, keeping only jets with $p_t > 3$ TeV and $|y| < 4$, We study the same groomers as for the Event2 study, as well as pruning and Y-pruning with $z_{\text{prune}} = 0.1$ and $f_{\text{prune}} = 0.5$.

Parton-Level Study The distributions obtained from Pythia and the analytic results from above are presented in Fig. 6.6. As for the case of the fixed-order studies in the previous section, the features observed in the parton-level simulation are very well reproduced by the analytic results, including the various transition

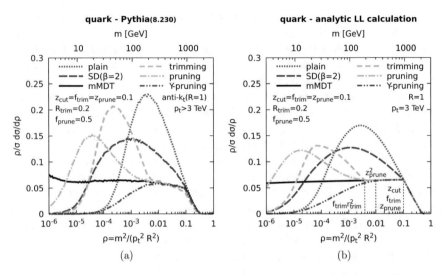

Fig. 6.6 Mass distribution obtained for the ungroomed jet (dotted, red) as well as with different groomers: SoftDrop($\beta = 2$) (long-dashed, blue), mMDT (solid, black), trimming (short-dashed, green), pruning (dot-dashed, cyan) and Y-pruning (dot-dot-dashed, magenta). The left plot (**a**) is the result of a Pythia parton-level simulation and the right plot (**b**) is the analytic results discussed in this chapter

points. The Pythia distributions tend to be more peaked than what is predicted from the analytic calculation, in particular in the regions where the distributions have a large double-logarithmic contribution. This effect would be (at least partially) captured by subleading, NLL, contributions, and in particular by contributions from multiple emissions which tend to increase the Sudakov and produce more peaked distributions. The latter should be present in the Pythia simulation but are absent from the above calculation.[8]

Finally, we see in Fig. 6.6 that for heavily-boosted jets, the transition points of trimming and pruning can be close to the electroweak scale. This is to keep in mind when using substructure techniques to tag boosted electroweak bosons.

Non-perturbative Corrections While the analytic calculations do a good job at reproducing the features observed in a parton-level Pythia simulation, the jet mass will also be affected by non-perturbative effects such as hadronisation and the UE. Ideally, we want these effects to be as small as possible to reduce the dependence on model-dependent, tuned, aspects of soft physics, which are not usually under good control and they can therefore obscure the partonic picture.

---

[8]They can easily be added to the ungroomed, SoftDrop and trimming calculations. We have not done it here because it clearly goes beyond the scope of these lecture notes.

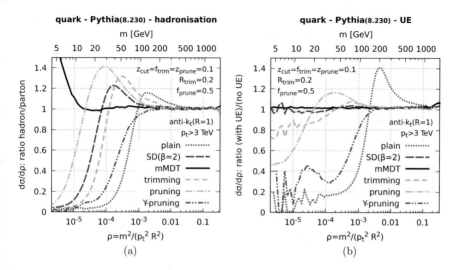

Fig. 6.7 Non-perturbative effects on the groomed jet mass distribution. The lines are as in Fig. 6.6. All results are obtained from Pythia8 simulations. The left plot (**a**) corresponds to hadronisation effects, i.e. the ratio of hadron-level to parton-level distributions. The right plot (**b**) shows the effects of the UE, i.e. the ratio of the mass distribution with UE effects on and off

We therefore switch on non-perturbative effects in Pythia8 and study how the reconstructed mass distributions are affected. Figure 6.7a shows the effects of hadronisation and it obtained by taking the ratio of the mass distribution with and without hadronisation effects. Figure 6.7b instead aims to study the impact of UE and it obtained by taking the ratio of the distribution with and without multiple-parton interactions (but with hadronisation).

Focusing first on UE effects, we clearly see the main idea behind grooming at play: by removing soft radiation at large angles, one significantly reduces the sensitivity to the UE, whereas the plain jet mass distribution shows a large distortion when this contribution is switched on. Furthermore, while all the groomers show almost no sensitivity to the UE at large mass ($\rho \gtrsim 0.002$ in Fig. 6.7b), differences start to appear at smaller masses. Y-pruning shows a relatively large sensitivity to the UE for $\rho \lesssim 0.002$, followed by pruning. This is likely due to UE effects on the plain jet mass affecting the determination of the pruning radius. Since the pruning radius will tend to be increased by UE effects, jets that would perturbatively be deemed as Y-pruning will fall in the I-pruning category once the UE is switched on. This is expectably the main source behind the drop observed in the Y-pruning curve in Fig. 6.7b. For the other groomers, trimming shows a smaller sensitivity, SoftDrop an even smaller one and the mMDT which is the most efficient at grooming away soft radiation shows almost no sensitivity to the UE.

This trend is similar when it comes to hadronisation corrections, Fig. 6.7a. While all the groomed jet mass distributions show a significantly smaller sensitivity to hadronisation than the plain jet mass distribution, one sees potentially sizeable effects at small values of ρ. As for the UE, Y-pruning shows the largest sensitivity amongst the groomers and mMDT clearly exhibits the smallest non-perturbative corrections.

Finally, by inspecting the mass scale on the upper horizontal axis, we note that for heavily boosted jets ($p_t = 3$ TeV in this case) it is worth keeping in mind that the non-perturbative effects can still be non-negligible around the electroweak scale.

Note finally that some degree of analytic control over the non-perturbative corrections to groomed jets can be achieved. This can be done either qualitatively by inspecting the expected non-perturbative scales to which each groomer is sensitive (see e.g. [123]), or more quantitatively using analytic models of hadronisation (see e.g. [79, 123, 180]).

6.4 Calculations for Signal Jets

Thus far, we have only discussed the case of QCD jets, which are initiated by high-energy quarks and gluons. Since the substructure tools discussed above are used extensively in the context of tagging boosted bosons—either as prong finders or as groomers—it is also interesting to discuss their behaviour for signal jets. Here, we will focus on electroweak bosons decaying to a quark-antiquark pair, leaving the more complicated case of the top quark aside. Our goal here is to give a very brief overview of how the tools discussed so far behave on signal jets. We will therefore only give analytic results at leading-order and rely mostly on Monte Carlo simulations to highlight the desired features associated with parton-shower and non-perturbative effects. Some degree of analytic calculation can be achieved for these effects as well but we will only highlight their main features here. More extensive analytic calculations, of both the perturbative and non-perturbative contributions, can be found in [158].

Zeroth-Order Behaviour At the lowest order in perturbation theory, we just have an electroweak boson decaying to a $q\bar{q}$ pair. When this two-parton system is passed to the groomer, the latter can either keep both partons in the groomed jet, in which case the jet is kept/tagged as a signal jet, or groom away one or both prongs in which case the jet is not tagger as a signal jet. In this simple situation, the signal efficiency—i.e. the fraction of signal jets kept after apply the jet substructure algorithm—is simply given by the rate of jets for which the two partons are kept by the groomer. This can be written as

$$\epsilon_S^{(\text{tagger})} = \int_0^1 dz \, P_X(z) \Theta^{(\text{tagger})}(z), \qquad (6.27)$$

where $P_X(z)$ is the probability that the electroweak boson X decays into a quark carrying a momentum fraction z of the boson and an anti-quark carrying a momentum fraction $1 - z$ of the boson. Crucially, the splitting function $P_X(z)$ does not exhibit the $1/z$ singularity at small z which we have encountered in the QCD case. This is nothing but our original argument that signal jets have a hard quark and a hard anti-quark, while QCD jets are dominated by a hard parton emitting soft gluons. Here, we will assume for simplicity a flat splitting probability $P_X(z) = 1$. This is correct for a heavily-boosted Higgs boson but only approximate for W and Z. For the latter, $P_{W/Z}(z)$ also depends on the polarisation of the boson. We refer the reader, for example, to the discussion in Section III.2.7 of [122] for a study of W polarisation in the context of jet substructure.

In Eq. (6.27), $\Theta^{(\text{tagger})}(z)$ denotes the action of the tagger on the $q\bar{q}$ pair. For a massive object X of mass m_X, the decay angle is given by $\theta^2 = \frac{m^2}{p_t^2 z(1-z)}$, or, again assuming that the angles are normalised to the jet radius R, $\theta^2 = \frac{\rho}{z(1-z)}$. The action of each tagger is then easy to write:

$$\Theta^{(\text{plain})}(z) = \Theta(\theta < 1),$$

$$\Theta^{(\text{mMDT})}(z) = \Theta(\theta < 1)\,\Theta(\min(z, 1-z) > z_{\text{cut}}),$$

$$\Theta^{(\text{SD})}(z) = \Theta(\theta < 1)\,\Theta(\min(z, 1-z) > z_{\text{cut}}\theta^\beta),$$

$$\Theta^{(\text{trim})}(z) = \Theta(\theta < 1)\,\Theta(\min(z, 1-z) > z_{\text{cut}} \text{ or } \theta < r_{\text{trim}}), \quad (6.28)$$

with pruning and Y-pruning showing the same behaviour as the mMDT at this order of the perturbation theory. Expressing θ as a function of z, we can rewrite all the above constraints as a cut on z and find (up to subleading power corrections in ρ)

$$\epsilon_S^{(\text{plain})}(z) = 1 - 2\rho,$$

$$\epsilon_S^{(\text{mMDT})}(z) = 1 - 2\max(\rho, z_{\text{cut}}),$$

$$\epsilon_S^{(\text{SD})}(z) = 1 - 2\max(\rho, z_{\text{cut}}(\rho/z_{\text{cut}})^{\beta/(2+\beta)}),$$

$$\epsilon_S^{(\text{trim})}(z) = 1 - 2\max(\rho, \min(f_{\text{trim}}, \rho/r_{\text{trim}}^2)). \quad (6.29)$$

These results show the same transition point as for the signal jet (at least at the lowest order of perturbation theory). Except at low p_t (or large mass), the mMDT (and (Y-)pruning) have a ρ-independent behaviour, with $\epsilon_S = 1 - 2z_{\text{cut}}$; the other taggers/groomers have an efficiency going asymptotically to 1 like a power of ρ, although in the case of trimming, this only happens at very small ρ, $\rho \ll z_{\text{cut}}r_{\text{trim}}^2$.

We can compare these results to Monte Carlo simulations. For simplicity, we use the Pythia8 generator, simulating the associated production of a Higgs and a Z boson, where the latter decays into (invisible) neutrinos and the Higgs boson decays to a $b\bar{b}$ pair. We reconstruct the jets using the anti-k_t algorithm with $R = 1$ and select the hardest jet in the event, imposing a cut on the jet p_t. The jet is then tagged/groomed and we deem the jet as tagged if the jet mass after grooming is within $\delta M = 20$ GeV of the Higgs mass, i.e. between 105 and 145 GeV, with

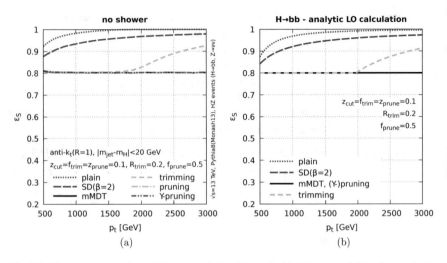

Fig. 6.8 Higgs reconstruction efficiency as obtained from Pythia8 (**a**) and a LO analytic calculation (**b**). The Pythia8 simulation is done at parton level with both the initial-state and final-state shower switched off. Different curves correspond to different taggers (see e.g. Fig. 6.6 for details)

$m_H = 125$ GeV. We study the Higgs tagging efficiency as a function of the p_t cut applied to the initial jet.

To compare to the analytic results, Eq. (6.29), we simulate parton-level results switching off both the initial and final-state showers. Results are presented in Fig. 6.8 (left) together with our simple analytic results (right). The analytic results capture very well the behaviour observed in the Monte Carlo simulations. In particular, all the features discussed above can be observed: the mMDT and (Y-)pruning remain constant as a function of the jet p_t and the efficiency of the other taggers/groomers increases with p_t, with the plain jet efficiency increasing more rapidly than the SoftDrop one. With our choices of parameters, the transition $\rho = z_{\text{cut}}$ (or f_{trim}) corresponds to a $p_t \approx 400$ GeV and is thus not visible on the plot. For trimming one sees the transition between the region dominated by the $z > f_{\text{trim}}$ condition at lower p_t and the region dominated by the $\theta < r_{\text{trim}}$ condition at larger p_t. The transition between the two regions happens at $p_t = m_H/(R_{\text{trim}}\sqrt{f_{\text{trim}}}) \approx 2$ TeV, in agreement with what is observed on the plot.

Final-State Radiation We now move to consider the effects of final-state radiation (FSR) on signal efficiency. The final-state gluons radiated by the $q\bar{q}$ pair can be groomed away, resulting in a decrease of the reconstructed jet mass. The jet mass can therefore fall below our lower cut $m_H - \delta M$ on the mass meaning that FSR is expected to reduce the signal efficiency. We know from our discussion of QCD jets in the previous sections that the emissions in a final-state shower can have logarithmically-enhanced effects on jet substructure observable. From an analytic viewpoint, these emissions would then have to be resummed to all orders.

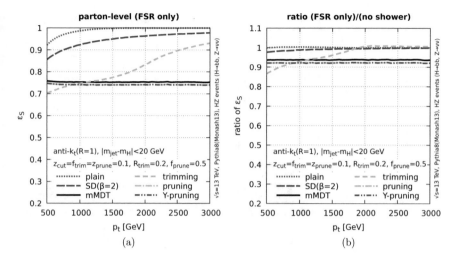

Fig. 6.9 Higgs reconstruction efficiency as obtained from Pythia8. (**a**) The simulation is done at parton level, including only final-state radiation. (**b**) The right plot shows the effects of final-state shower, i.e. the ratio to the efficiencies obtained with no final-state shower. See Fig. 6.8 for other details

While in practice it would be insightful to first consider the $\mathcal{O}(\alpha_s)$ case where a single gluon is emitted by the $q\bar{q}$ pair—similarly to what was done for the one-gluon emission case for QCD jets at LO—we directly turn to the situation where we include the full parton shower. We first discuss the case of final-state radiation—by the $q\bar{q}$ pair—and discuss initial-state radiation below. We therefore run Pythia8 simulations, still at parton level, but this time including final-state shower (and with the initial-state shower still disabled). The resulting efficiencies are plotted in Fig. 6.9. If one focuses on the right-hand plot, showing the ratio of the efficiencies obtained with final-state radiation to the efficiencies obtained without, we see a relatively small effect of FSR for all the substructure algorithms, even very small for the plain jet and SoftDrop. This is not true for trimming, for which the effect of FSR is to a large extend constant in p_t. In the case of trimming, we see that at small p_t, more precisely for $p_t < m_H/(R_{\text{trim}}\sqrt{z_{\text{cut}}}) \approx 2$ TeV, i.e. $\rho > z_{\text{cut}}r_{\text{trim}}^2$, the effect of FSR increases when decreasing p_t.

From an analytic perspective, the emission of FSR gluons can come with an enhancement proportional to $\log(\delta M^2/M_H^2)$ for a small-width mass window, or a logarithm of z_{cut}, f_{trim} or z_{prune}, all associated with soft gluon emissions. This is what drives the p_t independent loss of signal efficiency in the case of mMDT and (Y-)pruning in Fig. 6.9. For the plain jet and SoftDrop, this effect becomes suppressed by a power of M_H/p_t. Furthermore, in the case of trimming, due to the fixed trimming radius, the effect of final-state radiation is also enhanced by collinear logarithms of ρ/r_{trim}^2 for $r_{\text{trim}}^2 \ll \rho \ll 1$, i.e. in the intermediate p_t region. This logarithmically-enhanced effect is the main reason for the slow rise of the trimming signal efficiency between 500 GeV and 2 TeV.

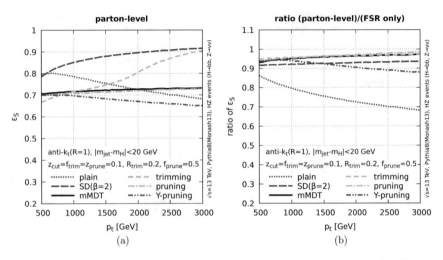

Fig. 6.10 (**a**) Higgs reconstruction efficiency obtained with Pythia8 at parton level. (**b**) effects of initial-state radiation, i.e. ratio to the efficiencies obtained with only final-state shower. See Fig. 6.8 for other details

Initial-State Radiation Next, we discuss the effect of initial-state radiation (ISR). Compared to the case of FSR, capturing an ISR gluon in the (groomed) jet shifts its mass up, meaning that it can go above $M_H + \delta M$, again lowering the efficiency. This effect is again potentially enhanced by a logarithm of δM^2. The results of our Monte Carlo study of ISR effects is presented in Fig. 6.10, where we a small effect for mMDT, SoftDrop, trimming and pruning, a slightly larger effect for Y-pruning and a sizeable loss of efficiency in the case of the plain jet.

In the case of the plain jet mass, one does get an enhancement of ISR effects by a logarithm of $M_H \delta M/p_t^2$, responsible for the loss of signal efficiency when increasing p_t. For groomed jets, one can show (see [158]) that this logarithm is typically suppressed by a power of M_H/p_t (related to the fact that the groomed jet radius decreases with p_t) and is replaced by a less harmful logarithm of z_{cut}, f_{trim} or z_{prune} coming from situations where a large-angle ISR gluon passes the grooming condition. The case of Y-pruning is a bit more complex as even when an ISR emission fails the pruning condition, it could have still affected (increased) the pruning radius and cause the Y-pruning condition to fail. This is the main source of the decrease of the signal efficiency observed for Y-pruning at large p_t in Fig. 6.10.

Non-perturbative Effects The effects of hadronisation and of the UE are presented in Figs. 6.11 and 6.12, respectively. Hadronisation corrections are generally small, especially for groomed jets where they are almost negligible. In the case of the plain jet, hadronisation effects tend to increase at large p_t but the correction remains within 10%. The case of UE corrections is more striking: the signal efficiency in the case of the plain jet is severely affected by UE contamination. After grooming, the UE correction becomes very small across the whole range of p_t studied.

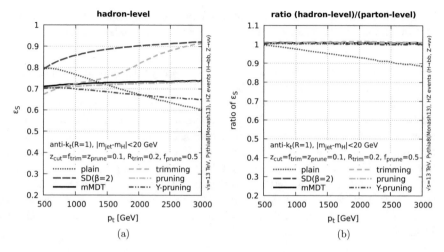

Fig. 6.11 (**a**) Higgs reconstruction efficiency obtained from Pythia8 at hadron level. (**b**) hadronisation effects, i.e. ratio to parton-level efficiencies. See Fig. 6.8 for details

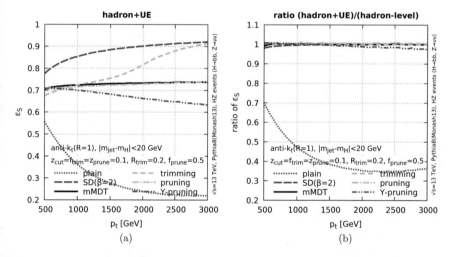

Fig. 6.12 (**a**) Higgs tagging efficiency obtained from a full Pythia8 simulation. (**b**) UE effects, i.e. ratio to efficiencies with UE switched off. See Fig. 6.8 for details

This is directly related to the initial idea behind grooming, namely to reduce soft contamination—and hence UE effects—by removing soft and large-angle emissions in the jet.

Once all effects are taken into account, the efficiency for groomed jets is found to be close to the initial prediction at leading order, with small corrections from ISR, FSR and non-perturbative effects. Trimming has a small extra p_t dependence at intermediate p_t coming from final-state radiation, and Y-pruning has a small loss of signal efficiency at large p_t due to initial-state radiation. This picture is contrasted by what happens in the case of the plain jet where ISR and, in particular, the UE have a sizeable effect, and hadronisation corrections are larger than for groomed jets. A consequence of this resilience of groomed jets is that, despite the smaller signal efficiency at leading-order, cf. Fig. 6.8, the groomed jet signal efficiency is clearly larger than the ungroomed signal efficiency once all effects beyond LO are included.

Chapter 7
Quark/Gluon Discrimination

In this chapter we discuss the application of jet substructure tools for discriminating between quark- and gluon-initiated jets. Before digging into the substructure aspects of the matter, let us briefly mention that there are many ways to define what a "quark jet" or a "gluon jet" is. Several possibilities are listed in Fig. 7.1. Amongst these possible definitions, many are clearly pathological, simply because a parton is not a physically well-defined object (cf. also our discussion about jets in Chap. 3). What is well-defined is a measurable quantity, that one can associate (in an inevitably ambiguous way) to an enriched sample of quarks or gluons. For simplicity, we often rely on event samples involving hard quarks or gluon in the Born-level process, but one has to be aware that this is not unambiguously defined approach and keep this in mind when interpreting the results. This is what we have already done in the previous chapter when generating $qq \rightarrow qq$ Pythia8 events as a proxy for quark jets and this is again what we will do here. Note that the better-defined definition in Fig. 7.1 depends on which sample is used. An investigation of this dependence can be found in [182].

That said, several processes one wants to measure at the LHC, like Higgs production through vector-boson-fusion, or new-physics events, such as cascades of supersymmetric particles, tend to produce quark jets while QCD backgrounds are gluon-dominated. This motivates the use of substructure tools to try and discriminate between the two. Some years ago, a wide range of discriminants has been systematically studied and compared [135]. It is not our goal to go through all the details of this study. Instead, we have selected a few representative discriminators and discussed their performance and their basic analytic properties. We focus on two main categories of tools: jet shapes, namely angularities and energy-correlation functions, and multiplicity-based observables, namely the iterated SoftDrop multiplicity. We conclude this chapter with a comparison of their performance (in the sense of Sect. 5.2) using Monte Carlo simulations.

Our Monte Carlo studies use Pythia8 (with the Monash13 tune). We generate "quark-initiate jets" using $qg \rightarrow Zq$ hard matrix elements and "gluon-initiated jets"

© Springer Nature Switzerland AG 2019
S. Marzani et al., *Looking Inside Jets*, Lecture Notes in Physics 958,
https://doi.org/10.1007/978-3-030-15709-8_7

What is a Quark Jet?
From lunch/dinner discussions

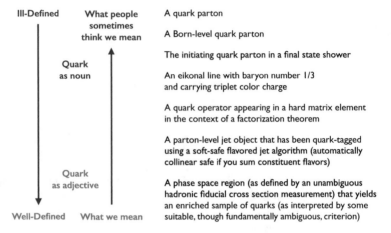

Fig. 7.1 Possible definitions of a "quark jet" or a "gluon jet" (from Ref. [136], see also [181])

using $q\bar{q} \rightarrow Zg$ events. In both cases, the Z boson is made to decay into invisible neutrinos and we focus on the hardest anti-$k_t(R = 0.5)$ jet in the event requiring $p_t > 500$ GeV.

7.1   Angularities, ECFs and Casimir Scaling

The motivation behind using jet shapes for quark-gluon discrimination is the observation that gluons tend to radiate more than quarks and jet shapes are precisely a measure of this radiation. Typical examples of shapes that can be used in this context are the angularities λ_α and the energy-correlation functions (ECFs) $e_2^{(\alpha)}$, introduced in Sect. 5.4. In both cases, one would expect a larger value of $v = \lambda_\alpha, e_2^{(\alpha)}$ for gluon jets than for quark jets and one can build an enhanced quark sample by simply imposing a cut $v < v_{\text{cut}}$.

We will first perform some analytic calculations for angularities and ECFs, before discussing their performance as quark-gluon separators. We will come back to this in Sect. 7.3, where we also discuss their robustness against non-perturbative effects.

Analytic Behaviour   For the purpose of the physics discussion we want to have, we will need a resummed calculation at NLL accuracy. At this accuracy, angularities and ECFs have the same structure, provided one uses a recoil-insensitive jet axis definition for angularities with $\alpha \leq 1$. This is easy to explain from a simple one-gluon emission argument (cf. e.g. Fig. 6.1). If θ denotes the angle between the emitted soft gluon and the recoiling hard parton, a standard four-vector recombination scheme, e.g. the E-scheme, would give an angle $(1 - z)\theta$ between the soft gluon and the jet axis and an angle $z\theta$ between the recoiling hard parton and the jet axis. This gives

$$\lambda_\alpha^{(\text{E-scheme})} = z[(1 - z)\theta]^\alpha + (1 - z)[z\theta]^\alpha = [z(1 - z)^\alpha + (1 - z)z^\alpha]\theta^\alpha, \qquad (7.1)$$

where the first contribution comes from the soft gluon and the second from the recoiling parton. For $\alpha = 1$, both partons contribute equally to give $\lambda_1^{(\text{E-scheme})} = 2z(1 - z)\theta \approx 2z\theta$. This leaves the LL behaviour unaffected but introduces recoil effects at NLL (with a resummation structure more complex than the simple exponentiation in (4.23). For $\alpha < 1$, $\lambda_\alpha^{(\text{E-scheme})} \approx z^\alpha\theta^\alpha$ dominated by the recoil of the hard parton, so recoil effects are already present at LL. If we use the winner-takes-all (WTA) axis—what we did in practice in our Monte Carlo simulations—angularities become recoil-free and we have

$$\lambda_\alpha^{(\text{WTA})} = z\theta^\alpha. \qquad (7.2)$$

This effect is not present for ECFs for which we have $e_2^\alpha = z(1 - z)\theta^\alpha \overset{z\ll 1}{\approx} z\theta^\alpha$, independently of the recombination scheme.

For $\alpha = 2$, angularities and ECFs are essentially equivalent to the mass—more precisely $m^2/(p_t R)^2$—and we can reuse the same results as in Chap. 4. These results can almost trivially be extended to a generic value of the angular exponent α. First, we need expressions for the radiators valid at NLL. This requires including the two-loop running-coupling corrections in the CMW scheme (see the discussion before Eq. (4.19)). For the plain jet, one finds a generalisation of Eqs. (4.20) and (4.21):

$$R_{\text{plain}}^{(\text{NLL})}(v) = \frac{C_i}{2\pi\alpha_s\beta_0^2}\left\{\left[\frac{1}{\alpha - 1}W(1 - \lambda) - \frac{\alpha}{\alpha - 1}W(1 - \lambda_1) + W(1 - \lambda_B)\right]\right.$$

$$\qquad\qquad (7.3)$$

$$+ \frac{\alpha_s\beta_1}{\beta_0}\left[\frac{1}{\alpha - 1}V(1 - \lambda) - \frac{\alpha}{\alpha - 1}V(1 - \lambda_1) + V(1 - \lambda_B)\right]$$

$$- \frac{\alpha_s K}{2\pi}\left.\left[\frac{1}{\alpha - 1}\log(1 - \lambda) - \frac{\alpha}{\alpha - 1}\log(1 - \lambda_1) + \log(1 - \lambda_B)\right]\right\},$$

where $W(x) = x \log(x)$, $V(x) = \frac{1}{2} \log^2(x) + \log(x)$ and we have introduced

$$\lambda = 2\alpha_s \beta_0 \log(1/v), \qquad \lambda_B = -2\alpha_s \beta_0 B_i, \qquad \text{and} \quad \lambda_1 = \frac{\lambda + (\alpha - 1)\lambda_B}{\alpha}.$$

$$(7.4)$$

Before discussing these results, let us point out that one can also apply grooming to the jet, using mMDT or SoftDrop, and compute the shape on the groomed jet. In this case, we get the same as Eq. (7.3) for $v > z_{\text{cut}}$ and a generalisation of Eq. (6.10) for $v < z_{\text{cut}}$:

$$R_{\text{mMDT/SD}}^{(\text{NLL})}(v) = \frac{C_i}{2\pi\alpha_s\beta_0^2} \left\{ \left[\frac{(\alpha + \beta)W(1 - \lambda_2)}{(\beta + 1)(\alpha - 1)} - \frac{\alpha W(1 - \lambda_1)}{\alpha - 1} \right. \right. \tag{7.5}$$

$$\left. - \frac{W(1 - \lambda_c)}{\beta + 1} + W(1 - \lambda_B) \right] + \frac{\alpha_s\beta_1}{\beta_0}$$

$$\left[\frac{(\alpha + \beta)V(1 - \lambda_2)}{(\beta + 1)(\alpha - 1)} - \frac{\alpha V(1 - \lambda_1)}{\alpha - 1} \right.$$

$$\left. - \frac{V(1 - \lambda_c)}{\beta + 1} + V(1 - \lambda_B) \right] - \frac{\alpha_s K}{2\pi}$$

$$\left[\frac{(\alpha + \beta)\log(1 - \lambda_2)}{(\beta + 1)(\alpha - 1)} - \frac{\alpha \log(1 - \lambda_1)}{\alpha - 1} \right.$$

$$\left. \left. - \frac{\log(1 - \lambda_c)}{\beta + 1} + \log(1 - \lambda_B) \right] \right\},$$

with

$$\lambda_c = 2\alpha_s\beta_0 \log(1/z_{\text{cut}}), \qquad \text{and} \quad \lambda_2 = \frac{(\beta + 1)\lambda + (\alpha - 1)\lambda_c}{\alpha + \beta}. \tag{7.6}$$

These expressions require a few comments. First of all, Eq. (7.3), for $\alpha = 2$, slightly differs from Eqs. (4.20) and (4.21). The difference is in the treatment of the B term which corresponds to hard collinear splittings where, as in Chap. 6 (cf. (6.11)), we have inserted the contribution from hard-collinear splittings in the double-logarithmic terms (see also Appendix A for a discussion on how to do this in practice). One can also notice that the limit $\beta \to \infty$ of Eq. (7.5) gives back Eq. (7.3) as expected. Furthermore, taking $\alpha = 2$ in the mMDT/SoftDrop case and neglecting the two-loop corrections, one recovers Eq. (6.11). Finally, we note that, although the above results have factors of $\alpha - 1$ in the denominator, they are finite for $\alpha \to 1$ (corresponding to the specific case of broadening or girth for angularities).

Given the above radiators, we can compute the probability that the angularity (or ECF) has a value smaller than v, i.e. the cumulative distribution, at NLL:

$$\Sigma^{(\text{NLL})}(v) = \frac{e^{-R(v)-\gamma_E R'(v)}}{\Gamma(1 + R'(v))}, \qquad (7.7)$$

where the factor $e^{-\gamma_E R'(v)}/[\Gamma(1 + R'(v))]$ accounts for multiple emissions (cf. (4.22)), and $R'(v)$ is the derivative of $R(v)$ with respect to $\log(1/v)$. Since the multiple-emission correction is already subleading, R' in (7.7) can be computed from the LL terms in R and we get (again, keeping the B term only to guarantee an endpoint at $\log(v) = B_i$)[1]

$$R'_{\text{plain}}(v) = \frac{C_i}{\pi \beta_0} \frac{1}{\alpha - 1} \log\left(\frac{1 - \lambda_1}{1 - \lambda}\right), \qquad (7.8)$$

$$R'_{\text{mMDT/SD}}(v) = \frac{C_i}{\pi \beta_0} \frac{1}{\alpha - 1} \log\left(\frac{1 - \lambda_1}{1 - \lambda_2}\right). \qquad (7.9)$$

Note finally that while Eq. (7.7) is only correct in the small jet radius limit and should include soft wide-angle emissions and non-global logs to reach full NLL accuracy, Eq. (7.7) includes all the NLL contributions for Soft-Dropped angularities which are insensitive to soft wide-angle emissions.

Comparison to Monte Carlo A comparison between the above analytic predictions and parton-level Monte Carlo simulations are shown in Fig. 7.2, for different values of the angularity exponent for SoftDrop jets, and in Fig. 7.3, for different levels of grooming for $\lambda_{\alpha=1}$. Overall, we see that there is a good agreement between the analytic calculation and the Monte Carlo simulations. We recall that our resummed calculation should not be trusted in the region of large v where an exact fixed-order calculation would be needed. This could be obtained from NLO Monte Carlo generators like NLOJet++ [183] for dijet hard processes (here one would need a 3-jet NLO calculation for the angularity distribution) and MCFM [184–186] for W/Z+jet events (here we would need W/Z+2 jets at NLO for the angularity distribution). The NLO distributions could then be matched to the resummed calculation to obtain a final prediction which is valid at the same time in the resummation-dominated region (small angularity) and in the fixed-order-dominated region (large angularity). More importantly, Figs. 7.2 and 7.3 show the expected clear separation between the quark and gluon samples, with smaller values of the angularity for the quark jets.

---

[1]In practice, this definition of $R'_{\text{mMDT/SD}}(v)$ introduces a discontinuity in the differential distribution at $v = z_{\text{cut}}$. This discontinuity is strictly-speaking subleading and can be avoided by defining R' using a finite-difference derivative: $R'(v) = [R(ve^{-\Delta}) - R(v)]/\Delta$, with Δ a constant number, which respects NLL accuracy (see [50]). This is what we have done for the results presented below, using $\Delta = 0.5$.

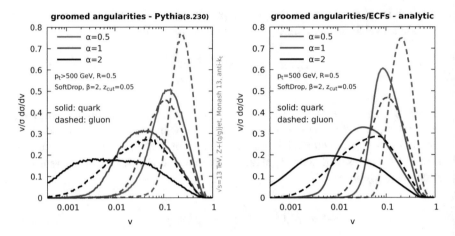

Fig. 7.2 Distribution of a sample of groomed angularities for quark (solid lines) and gluon (dashed lines) jets. The left plot corresponds to parton-level Pythia simulations and the right plot to the analytic results obtained in these lecture notes

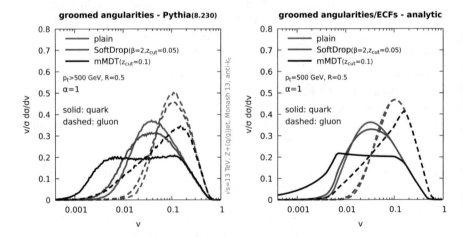

Fig. 7.3 Same as Fig. 7.2, this time for a fixed angularity λ_1, varying the groomer

Quark-Gluon Discrimination and Casimir Scaling With the above results at hand, we can finally discuss the performance of angularities and energy-correlation functions to separate quark jets from gluon jets. This is simply done by imposing a cut $v < v_{cut}$ on angularities or ECFs. On the analytic side, the quark and gluon efficiencies are therefore directly given by $\Sigma_{q,g}$ computed above. An interesting behaviour emerges from these analytic results. If we look at Eqs. (7.3) and (7.5) at leading-logarithmic accuracy, the only difference between quark and gluon jets is the colour factor—C_F for quarks and C_A for gluons, in from of the radiators. This means that we have

$$\epsilon_g \overset{\text{LL}}{=} (\epsilon_q)^{C_A/C_F}. \tag{7.10}$$

This relation is often referred to as *Casimir scaling* (see e.g. [143]). This means that the leading behaviour of quark-gluon tagging will follow Eq. (7.10) regardless of the angularity (of ECF) exponent and of the level of grooming.

Departures from Casimir scaling will start at NLL accuracy. In our collinear/small-R limit, these means that there can be three sources of Casimir-scaling violations: hard-collinear corrections (the B term), two-loop running-coupling corrections, and multiple-emissions (cf. Eq. (7.7)). Of these three effects, only the first and the last give scaling violations since two-loop running coupling corrections are also simply proportional to C_i. This is illustrated in Fig. 7.4a, where we see that the LL result gives perfect Casimir scaling, and the inclusion of the hard collinear splitting and the multiple-emission corrections both slightly increase the quark-gluon discrimination performance. The correction due to the B-term is proportional to $B_g - B_q$ which is small and positive. The effect of multiple emissions starts at $\mathcal{O}(\alpha_s^2)$ in the perturbative expansion and is proportional to $(C_A - C_F)$. In practice, this last effect appears to have the largest impact. A direct consequence of Casimir scaling is that the quark-gluon discriminative power remains relatively independent of the jet p_t as shown in Fig. 7.4b for both our analytic calculation (dashed lines) and Pythia8 parton-level simulations (solid lines).

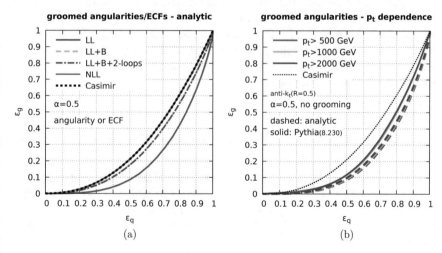

Fig. 7.4 (**a**) analytic predictions for the quark-gluon separation ROC curve using different approximations. (**b**) ROC curve for different values of p_t, shown for both Pythia8 simulations (solid) and our analytic calculation (dashed)

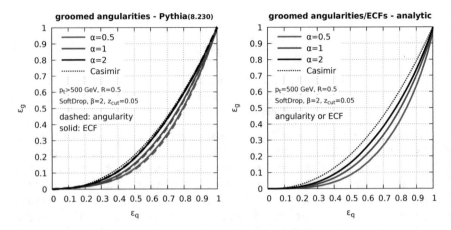

Fig. 7.5 Distribution of a sample of groomed angularities for quark (solid lines) and gluon (dashed lines) jets. The left plot corresponds to parton-level Pythia simulations and the right plot to the analytic results obtained in these lecture notes

All these effects are discussed at length in Ref. [143] and we refer the reader to this discussion for further details. ROC curves for quark-gluon discrimination are shown in Fig. 7.5 for both Pythia (at parton level) and our analytic calculation, for jets groomed with SoftDrop. We see a good level of agreement between the two although the analytic results tend to produce a slightly larger quark-gluon discrimination that Pythia. It is however notorious that different Monte Carlo generators tend to predict relatively different deviations from Casimir scaling, both at parton and hadron level. We refer to Ref. [181] for more details about this. Above all, we conclude from Fig. 7.5 that smaller values of α give better discrimination, with very similar results obtained for angularities and energy-correlation functions. We will come back to this in Sect. 8.3 when discussing the performance and robustness of quark-gluon discriminators.

7.2 Beyond Casimir Scaling with Iterated SoftDrop

Given the observation made in the previous section that angularities and energy-correlation functions produce quark-gluon discriminators which depart from Casimir scaling only due to subleading corrections, it is natural to wonder if it is possible to find substructure tools which have a different behaviour already at leading-logarithmic accuracy.

The behaviour one would want to obtain is a Poisson-like behaviour like what the particle multiplicity in a jet, or the charged-track multiplicity, typically achieve. In this section, we discuss a tool, namely the *Iterated SoftDrop (ISD) multiplicity* introduced in Sect. 5.4.4 and show that it achieves a Poisson-like behaviour already at LL while remaining infrared-and-collinear safe (contrary to particle or charged-track multiplicity). As above, we will first briefly discuss the analytic structure of ISD multiplicity and compare the resulting performance with Monte-Carlo simulations.

ISD Multiplicity at LL The main interesting features of ISD multiplicity already arise at leading logarithmic accuracy, so we will focus on this in what follows. The key observation is that at LL, all the emissions from the hard (leading) parton are soft and collinear, strongly ordered in angle and independent from one another. The fact that the emissions are independent automatically guarantees that, if ν is the probability that one emission is counted by the ISD de-clustering procedure, i.e. passes the SoftDrop condition, then the probability to have n emission passing the SoftDrop condition follows a Poisson distribution

$$\frac{1}{\sigma}\frac{d\sigma}{dn_{\mathrm{ISD}}} = e^{-\nu}\frac{\nu_{\mathrm{ISD}}^n}{n_{\mathrm{ISD}}!}.\tag{7.11}$$

We now need to compute ν explicitly. This is straightforward since, at LL, the probability to have an emission that passes the SoftDrop condition is simply given by (measuring angles in units of the jet radius as usual)

$$\nu = \int_0^1 \frac{d\theta^2}{\theta^2} dz\, P_i(z)\frac{\alpha_s(z\theta p_t R)}{2\pi}\Theta(z > z_{\mathrm{cut}}\theta^\beta).\tag{7.12}$$

For the ISD multiplicity to be IRC-safe, ν has to remain finite. This can easily be achieved by using a negative value for β, guaranteeing a finite phase-space for the emissions (cf. e.g. the Lund diagram of Fig. 6.2b). Alternatively, we can manually impose a minimum k_t cut, $z\theta > \kappa_{\mathrm{cut}}$, on the emissions which pass the SoftDrop condition, or stop the iterative de-clustering procedure at a minimum angle θ_{cut}.

These three options correspond to the three regions of the Lund diagram shown in Fig. 7.6. The corresponding analytic expressions for ν can be obtained exactly as for the radiators computed for angularities in the previous section (this time keeping

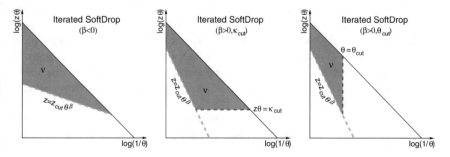

Fig. 7.6 Lund diagrams representing the regions in which Iterated SoftDrop counts the emissions. From left to right we have $\beta < 0$, $\beta > 0$ with a cut on k_t and $\beta > 0$ with an angular cut

only LL term, and hard collinear splittings). One finds (assuming $\kappa < z_{\text{cut}}$ for the second case)[2]:

$$
\nu_{\beta<0} = \frac{C_i}{2\pi\alpha_s\beta_0^2}\left[\frac{-1}{1+\beta}W(1-\lambda_c) - \frac{\beta}{1+\beta}W\left(1+\frac{\lambda_c}{\beta}\right)\right]
$$

$$
- \frac{C_i}{\pi\beta_0}\log\left(1+\frac{\lambda_c}{\beta}\right)B_i, \tag{7.13}
$$

$$
\nu_{\beta>0,\kappa} = \frac{C_i}{2\pi\alpha_s\beta_0^2(1+\beta)}\left[-W(1-\lambda_c) - (\lambda_c+\beta)\log(1-\lambda_\kappa)\right.
$$

$$
\left. - \lambda_c - \beta\lambda_\kappa\right] - \frac{C_i}{\pi\beta_0}\log(1-\lambda_\kappa)B_i, \tag{7.14}
$$

$$
\nu_{\beta>0,\theta} = \frac{C_i}{2\pi\alpha_s\beta_0^2(1+\beta)}\left[-W(1-\lambda_\theta) - \frac{W(1-\lambda_c)}{1+\beta}\right.
$$

$$
\left. + \frac{W(1-\lambda_c-(1+\beta)\lambda_\theta)}{1+\beta}\right] - \frac{C_i}{\pi\beta_0}\log(1-\lambda_\theta)B_i, \tag{7.15}
$$

with

$$
\lambda_c = 2\alpha_s\beta_0\log\left(\frac{1}{z_{\text{cut}}}\right), \quad \lambda_\kappa = 2\alpha_s\beta_0\log\left(\frac{1}{\kappa_{\text{cut}}}\right), \quad \text{and} \quad \lambda_\theta = 2\alpha_s\beta_0\log\left(\frac{1}{\theta_{\text{cut}}}\right),
$$

Counting logarithms of z_{cut}, κ_{cut} and θ_{cut}, all the above expressions show a double-logarithmic behaviour. An easy way to see this is to compute ν using a fixed-coupling approximation (equivalent to taking the limit $\beta_0 \to 0$ in the above results). For example, for the representative $\beta < 0$ case we will use in what follows, one has

$$
\nu_{\beta<0} \overset{\text{f.c.}}{=} \frac{\alpha_s C_i}{\pi}\frac{-1}{\beta}\left[\log^2\left(\frac{1}{z_{\text{cut}}}\right) + 2B_i\log\left(\frac{1}{z_{\text{cut}}}\right)\right]. \tag{7.16}
$$

Figure 7.7 shows the ISD multiplicity distributions for quark and gluon jets, obtained from (parton-level) Pythia8 simulations (left) and using the analytic expressions above (right). Each plot shows different values of z_{cut}. For these plots, we have used $\beta = -1$, corresponding to a cut on the relative k_t of the emissions. To make this more concrete, the value of z_{cut} is given as a function of the corresponding k_t cut. In the case of the Pythia8 simulations, the cut has been adapted using the p_t of each individual jets. Overall, we see that the analytic calculation captures the main features of the Monte Carlo simulation, albeit with distributions which tend to be peaked towards lower multiplicities than in Pythia8. We note that NLL corrections,

---

[2]These first two results can be directly derived from Eqs. (7.3) and (7.5). The third corresponds to the radiator for the SoftDrop grooming radius originally computed in Ref. [50] and discussed in Sect. 9.1 below.

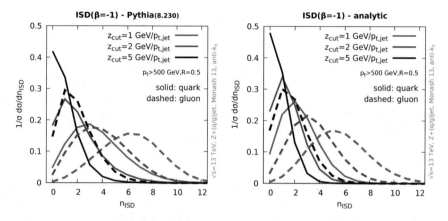

Fig. 7.7 Distribution of ISD multiplicity for $\beta = -1$, varying z_{cut}. The value of z_{cut} is given as a dimensionful k_t scale, normalised to $p_t R$. The left plot corresponds to parton-level Pythia simulations (for which z_{cut} is re-calculated for each jet) and the right plot to the analytic calculation, Eqs. (7.11) and (7.13). Solid lines correspond to quark jets, while dashed lines correspond to gluon jets

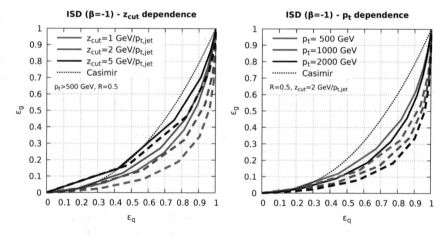

Fig. 7.8 Quark-gluon discrimination (ROC curve) using Iterated SoftDrop. The left plot uses a fixed jet p_t cut and varies the Iterated SoftDrop cut (defined as in Fig. 7.7). For the right plot, z_{cut} is fixed to $2\,\text{GeV}/(p_t R)$ and the cut on the jet p_t is varied

computed in the initial ISD study, Ref. [148], improves the agreement between the two. One particular effect that becomes relevant at NLL is that the flavour of the leading branch followed through the ISD declustering can change. This is included in Pythia8 via the DGLAP splitting functions and can be tracked analytically as well.

Quark-Gluon Discrimination The ROC curves obtained for quark-gluon discrimination are presented in Fig. 7.8, for Pythia (solid) and the LL analytic calculations (dashed). The left plot corresponds to the distributions shown in Fig. 7.7. First, we see that the discriminating power improves with lower z_{cut}. This is expected since the phase-space for emissions increases and so does v. Then, although the analytic calculation tends to over-estimate the discriminating power, the generic trend remains decently reproduced and we see, in particular, that the agreement is better at larger z_{cut} where the distribution is expected to have smaller non-perturbative corrections. It is worth pointing out that the flavour-changing effects briefly mentioned above and appearing at NLL accuracy would have the effect that quark and gluon jets would become more similar as we go to smaller angles, hence reducing the discriminating power.

Finally, the right plot of Fig. 7.8 shows that the discriminating power of ISD multiplicity improves at larger p_t (for a fixed k_t cut). This is again a consequence of the fact that the phase-space available for emissions, and hence v, increases. This contrasts with the angularities discussed previously: while the latter remain close to Casimir scaling at any energy, the performance of ISD multiplicity improves for larger jet p_t.

7.3   Performance and Robustness

To conclude this study of quark-gluon tagging, we compare several quark-gluon discriminators in terms of both their performance and their robustness. This is based on Pythia8 Monte Carlo simulations and we reiterate the caveat that the quark-gluon separation varies between Monte Carlo (cf. [136]), so this should be taken as a highlight of the main features rather than a full study. The main goal of this discussion is to stress more explicitly that, as introduced in Sect. 5.2, a high-quality substructure tool needs obviously to have a strong discriminating power, but at the same time it small sensitivity to non-perturbative effects is also desirable.

We first specify our quality measures for performance and robustness. For this, let us consider a given quark-gluon discriminator at a fixed working point (i.e. a given cut value). To treat quarks and gluons symmetrically, we define performance as the geometric mean of the quark significance and the gluon significance:

$$\Gamma_{sym} = \sqrt{\frac{\epsilon_q}{\sqrt{\epsilon_g}} \frac{1 - \epsilon_g}{\sqrt{1 - \epsilon_q}}}, \tag{7.17}$$

where one has used the fact that to tag gluon jets, one would impose a cut $v > v_{cut}$ and $\epsilon_{v > v_{cut}} = 1 - \epsilon_{v < v_{cut}}$. Robustness is then quantified through resilience, as introduced in Sect. 5.2, Eq. (5.1). For simplicity, we will focus here on the resilience against non-perturbative effects including both hadronisation and the Underlying Event (UE). These effects could be studied separately but this goes beyond the scope of this book. We note however that in our case, resilience is dominated by hadronisation effects, with UE having a much smaller impact. Finally, note that

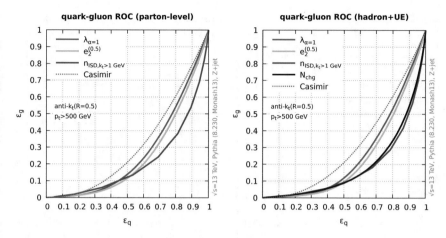

Fig. 7.9 ROC curves for a representative series of quark-gluon taggers: broadening, $\lambda_{\alpha=1}$ (red), energy-correlation function $e_2^{(\alpha=0.5)}$ (green), Iterated SoftDrop with $\beta = -1$ and $z_{\text{cut}} = 1 \text{ GeV}/p_{t,\text{jet}}$, and the charged track multiplicity. All the results are shown for Pythia8 simulations with a jet p_t cut of 500 GeV. The left plot corresponds to parton-level events while the right plot corresponds to full simulations including hadronisation and the Underlying Event. The charged-track multiplicity is not shown at parton level

both the performance Γ_{sym} and the resilience ζ can be computed for any fixed cut on a shape or multiplicity.

First, we compare the performance of a few representative tools discussed earlier in this section: girth or broadening, equivalent to the angularity $\lambda_{\alpha=1}$, energy-correlation function $e_2^{(\alpha=0.5)}$, the ISD multiplicity with $z_{\text{cut}} = 1 \text{ GeV}/p_{t,jet}$ (corresponding to a k_t cut of 1 GeV), and the charged track multiplicity. The ROC curves are shown in Fig. 7.9 for Pythia8 simulations at parton level (left) and at hadron level including the Underlying Event (right). At small quark efficiency ($\epsilon_q \lesssim 0.5$) angularities and energy correlation functions tend to give a better discriminating power. At larger quark efficiency multiplicity-based discriminators show a better performance, with the ISD and charged-track multiplicities behaving similarly.

We now discuss both the performance and resilience of our representative sample of quark-gluon taggers. This is first shown in Fig. 7.10 for the full ROC curves corresponding to Fig. 7.9, i.e. where the lines are obtained by varying the cut on the shape or multiplicity. The empty symbols correspond to a fixed quark efficiency of 0.5 at hadron+UE level, while the solid symbols correspond to a symmetric working point where $\epsilon_q = 1 - \epsilon_g$ (at hadron+UE level).[3] The charged-track multiplicity is not plotted simply because it is not well-defined at parton level.

---

[3]For multiplicity-based observables, we have interpolated linearly between the discrete multiplicities.

Fig. 7.10 Quark-gluon tagging quality: performance v. resilience for the taggers used in Fig. 7.9. The curves correspond to varying the cut on the jet shape or multiplicity. Solid (empty) points correspond to the specific working point for which $\epsilon_q = 1 - \epsilon_g$ ($\epsilon_q = 0.5$). Performance is computed at hadron+UE level and resilience includes both hadronisation and UE effects

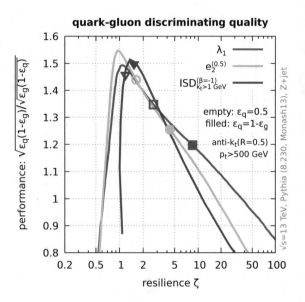

We see that angularities and ECFs give their best performance at relatively low quark efficiency, corresponding to a fairly low resilience. As the quark efficiency decreases (going to $\epsilon_q = 0.5$, then $\epsilon_q = 1 - \epsilon_g$) performance decreases but one gains resilience. A similar behaviour is seen for ISD although the highest performance is observed for larger quark efficiencies and large resilience at yet larger quark efficiencies. For our 500-GeV sample, the best performance is achieved by ECF($\alpha = 0.5$) closely followed by ISD, with the latter showing a slightly better resilience against non-perturbative effects. At lower Γ_{sym} this is inverted, with shape-based variables becoming more resilient than ISD.

The crucial observation one draws from Fig. 7.10 is that, generally speaking, there is a trade-off between performance and resilience. This pattern is seen repeatedly in substructure studies (we will see another example in our two-prong-tagger study in the next chapter) and can be understood in the following way: tagging constrains patterns of radiation inside a jet; usually, increasing the phase-space over which we include the radiation, and in particular the region of soft emissions, means increasing the information one includes in the tagger and hence increasing the performance; at the same time, the region of soft emissions being the one which is most sensitive to hadronisation and the Underlying Event, one also reduces resilience.

To finish this study of quark-gluon taggers, we show in Fig. 7.11 how the quark-gluon tagging quality varies with the jet p_t. From small to big symbols, we have used $p_t > 500$ GeV, $p_t > 1$ TeV and $p_t > 2$ TeV, and we have focused on the point for which $\epsilon_q = 1 - \epsilon_g$. The left plot shows this for two different choices of parameters (two exponents for angularities and ECFs and two z_{cut} for ISD). We see clearly that, as expected from our earlier studies, the performance of ISD

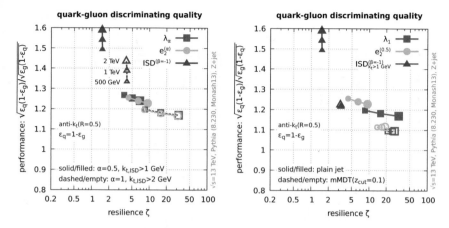

Fig. 7.11 Plot of performance v. resilience for quark-gluon taggers, as in Fig. 7.10, varying the cut on the jet p_t, using the working point $\epsilon_q = 1 - \epsilon_g$. The left plot shows two different choices of parameters for the taggers. The right plot shows two different levels of grooming

increases with the jet p_t while that of shape-based taggers remains roughly constant. Conversely, shape-based taggers become more resilient at larger p_t, highlighting again a trade-off between performance and resilience.

The right plot of Fig. 7.11 shows two different levels of grooming: the plain jet and a jet groomed with mMDT.[4] The dependence on the jet p_t is the same as what was already observed for the left plot (although, for mMDT jets, the performance of ISD only increases marginally). What is more interesting is that one clearly sees that grooming has the effect of reducing the performance and increasing the resilience. Since grooming is (almost by definition) removing soft emissions at large angles, this is another textbook example of a trade-off between performance and resilience. We note however that these conclusions are relatively sensitive to the choice of working point. For example, Fig. 7.12 shows the same result as Fig. 7.11 but now selecting for each method the working point which maximises performance. In this case, we see that all methods give similar results both in terms of performance and in terms of resilience, with even a small preference for ECFs (with $\alpha = 0.5$) if one is looking for sheer performance. It is worth pointing out that in this case the quark and gluon efficiencies are relatively low, meaning that (1) one might be affected by issues related to lower statistics and (2) we are in a region where the discreteness of ISD can have large effects which one would need to address in a more complete study.[5]

---

[4]In the case of ISD, we have applied mMDT recursively, giving a behaviour equivalent to using $\beta = 0$ and a k_t cut as shown in the middle plot of Fig. 7.6.

[5]For the results of Fig. 7.12 we have simply interpolated between different points in the distribution.

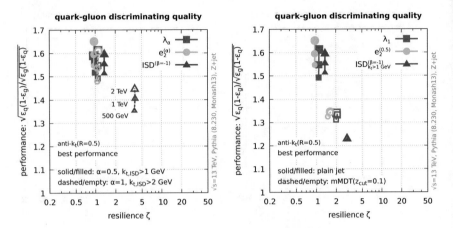

Fig. 7.12 Plot of performance v. resilience for quark-gluon taggers, as in Fig. 7.11 but now using the working point that maximises performance for each setup

As a final comment, we point out that, given the different behaviours seen between shape-based taggers and multiplicity-based taggers, it would be interesting to study their combination in a multivariate analysis. It would also be interesting to see how recent quark-gluon taggers based on deep learning techniques use the information relevant for ECFs and ISD.

Chapter 8
Two-prong Tagging with Jet Shapes

Two-prong taggers aim at discriminating massive objects that decay into two hard QCD partons (usually quarks), from the background of QCD jets. This signal is often an electroweak boson (H/W/Z) but it can also be a new particle (see Chap. 10 for examples).

Our goal in this chapter is two-folded and it closely follows what was done in the previous chapter for quark-gluon tagging. First, we want to give a brief insight into analytic properties of two-prong taggers, mainly selecting a few representative substructure tools and comparing their behaviour in Monte Carlo simulations with analytic results. Then, we will perform a comparative Monte Carlo study of the taggers discriminating properties, assessing both their performance and their resilience against non-perturbative effects.

8.1 A Dive into Analytic Properties

Two-prong taggers used for Run-II of the LHC tend to combine two major ingredients: a two-prong finder also acting as a groomer, and a cut on a jet shape for radiation constraint. Since groomers have already been extensively discussed in Chap. 6, in this chapter we are going to focus on the understanding of jet shapes and of their interplay with grooming. Note that a variety of jet shapes can be used in the context of tagging two-pronged boosted objects: Y-splitter, N-subjettiness, ECFs, pull, and so on. We will only select a few for our discussion.

While computations for groomers and prong-finders, such as the modified MassDrop Tagger or SoftDrop, have seen a lot of development towards precision calculations in the last few years and one can say that they are under good analytic control, the situation for jet shapes is more complex. This can be understood as follows: imagine one wants to tag a boosted object around a mass M_X; one would typically first require that the jet mass (groomed or ungroomed) is in a window

© Springer Nature Switzerland AG 2019
S. Marzani et al., *Looking Inside Jets*, Lecture Notes in Physics 958,
https://doi.org/10.1007/978-3-030-15709-8_8

close to M_X and then that the cut on the jet shape is satisfied; for QCD jets, which constitute the background, this means that we need to consider at least two emissions inside the jet—one setting the jet mass, the second setting the value of the shape—so calculations for the QCD background will start at order α_s^2 in the perturbative expansion, compared to α_s for groomers or quark/gluon taggers. That said, calculations now exist for a range of jet shapes (see e.g. [143, 157, 187–190]), noticeably ECFs and N-subjettiness, in both the direct QCD approach used in this book and in SCET.

To keep the discussion simple, we will assume that, on top of working in the boosted limit $m \ll p_{t,\text{jet}}$, the cut on the jet shape, $v < v_{\text{cut}}$, is also small so we can study the effect of the shape in the leading-logarithmic approximation. Technically, since we expect signal jets to mostly exhibit small values of v—i.e. there is less radiation in a signal jet than in QCD background jets—this approximation seems reasonable. For practical phenomenological applications however cuts on jet shapes are not much smaller than one and so finite v corrections are potentially sizeable. The leading-logarithmic approximation we will adopt in what follows, treating logarithms of $m/p_{t,\text{jet}}$ and v_{cut} (and, optionally of the grooming z_{cut} parameter) on an equal footing, is nevertheless sufficient to capture the main properties of two-prong taggers and differences between them.

For the purpose of this book, we will focus on three different shapes: the N-subjettiness ratio τ_{21}, with $\beta = 2$, which has a fairly simple structure and has been used at the LHC (albeit with $\beta = 1$). We will then move to the dichroic version of the τ_{21} ratio (see Eq. (5.30)) in order to illustrate how separating the grooming and prong-finding parts of the tagger could be helpful. Finally, we will discuss the ECFs $C_2^{(\beta=2)}$ and $D_2^{(\beta=2)}$. The latter in particular shows a very good discriminating power and it is used at the LHC (albeit with $\beta = 1$).

A typical LL calculation involves two steps: (1) compute an expression for the shape valid at LL and (2) use it to derive an expression for the mass distribution with a cut on the jet shape, or the distribution of the shape itself. The calculations for QCD jets will be followed by a calculation for signal (W/Z/H) jets and a comparison to Monte Carlo simulations done using the Pythia8 generator. Note that the analytic calculations below focus on computing the jet mass distribution imposing a cut on the jet shape: $(\rho/\sigma \, d\sigma/d\rho)_{v<v_{\text{cut}}}$. We can deduce the cumulative and differential distribution for the shape itself:

$$\Sigma(v) = \frac{(d\sigma/d\rho)_{v<v_{\text{cut}}}}{(d\sigma/d\rho)_{\text{no cut}}} \quad \text{and} \quad \frac{v \, d\sigma}{\sigma \, dv} = v\frac{d\Sigma}{dv}. \quad (8.1)$$

The background efficiency in a given mass window can also be obtained from the mass distribution with a cut on the shape via

$$\epsilon_B(\rho_{\text{min}}, \rho_{\text{max}}; v_{\text{cut}}) = \int_{\rho_{\text{min}}}^{\rho_{\text{max}}} d\rho \left. \frac{d\sigma}{d\rho}\right|_{v<v_{\text{cut}}}. \quad (8.2)$$

8.1.1 N-subjettiness $\tau_{21}^{(\beta=2)}$ Ratio

Approximate τ_{21} Value at LL To fully specify the definition of the τ_{21} ratio we are working with, it is not sufficient to specify the value of the β parameter, one also needs to specify the choice of axes. For our choice of $\beta = 2$, it is appropriate to work either with minimal axes, i.e. the axes that minimise the value of τ_N, or exclusive generalised-k_t axes with $p = 1/\beta = 1/2$. Let us consider a set of n emissions. For the purpose of our LL calculation, we can assume that they are strongly ordered in "mass" (or to be more precise in their contribution to the mass) i.e. $\rho_1 \gg \rho_2 \gg \cdots \gg \rho_n$, with $\rho_i = z_i\theta_i^2$, and strongly ordered in energy and angle (i.e., for example, $\theta_i \gg \theta_j$ or $\theta_i \ll \theta_j$ for any two emissions i and j). For the sake of definiteness, let us work with axes defined using the generalised-k_t ($p = 1/2$) exclusive subjets. We should thus first go through how our set of emissions is clustered. The generalised-k_t clustering will proceed by identifying the smallest $d_{ij} = \min(z_i, z_j)\theta_{ij}^2$ distance. Using $i = 0$ to denote the leading parton and assuming $z_i \ll z_j$, we have

$$d_{i0} = z_i\theta_i^2 = \rho_i, \tag{8.3}$$

$$d_{ij} = z_i\theta_{ij}^2 \approx z_i \max(\theta_i^2, \theta_j^2) \geq z_i\theta_i^2 \equiv \rho_i. \tag{8.4}$$

The overall minimal distance will therefore be the smallest of the ρ_i's, i.e. ρ_n. This can be realised in two ways: either the distance between emission n and the leading parton ($d_{n0} = \rho_n$) of the distance between emission n and any emission k with $\theta_k \ll \theta_n$ (for which Eq. (8.4) gives $d_{nk} \approx \rho_n$). In the second case, we also have $z_k \gg z_n$. Due to the energy ordering—and the fact that for $\beta = 2$ recoil effects can be neglected—after clustering particle n with either the leading parton or emission k, one gets a situation with the leading parton and emissions $1, \ldots, n-1$. The above argument can then be repeated, clustering particles $n - 1, n - 2, \ldots, 2, 1$ successively. This means that the τ_1 axis will be the jet axis—equivalent to the leading parton in this case—and the two exclusive generalise-k_t axes used for τ_2 will be aligned with the leading parton and with the largest ρ_i emission, i.e. with emission 1.[1]

With these axes, it is easy to deduce the value of τ_1 and τ_2 for our set of emissions:

$$\tau_1 = \sum_{i=1}^{n} z_i\theta_i^2 = \rho \approx \rho_1, \tag{8.5}$$

$$\tau_2 = \sum_{i=1}^{n} z_i \min(\theta_i^2, \theta_{i1}^2) \approx \rho_2, \tag{8.6}$$

---

[1]The argument can be extended to the N exclusive axes used for τ_N which would be aligned with the leading parton and with emissions $1, \ldots, N - 1$.

where, in the second line, the contribution from emission 1 vanishes.

Note that the above derivation is slightly incomplete: on top of the n emissions from the leading parton, we can also have secondary emissions from the leading emissions $1, \ldots, n$, i.e., in our angular-ordered limit, emissions "j" from the leading parton i with $z_j \ll z_i$ and $\theta_{ij} \ll \theta_i$. These will not affect the finding of the two axes needed to compute τ_2 but secondary emissions from emission 1 can dominate τ_2. Specifically, an emission with a momentum fraction z_2 relative to z_1 emitted at an angle θ_{21} from emission 1 would give

$$\tau_{2,\text{secondary}} \approx z_1 z_2 \theta_{12}^2 \quad \text{i.e.} \quad \tau_{21,\text{secondary}} \approx z_2 \frac{\theta_{12}^2}{\theta_1^2}. \tag{8.7}$$

Another way to view this is to consider that the two axes used to compute τ_2 define a partition of the jet in two subjets (one around the leading parton, the second around emission 1). The total τ_2 is therefore the sum of the individual contributions from these two subjets, i.e. from the sum of $z_i \theta_{i,\text{axis}}^2$ in these two subjets and the dominant contribution can come from either subjet. This is in contrast with all the calculations done previously in this book, which were only sensitive to primary emissions. It should however not come as a surprise since we are discussing tools which measure the radiation pattern around a two-prong structure so one should expect a contribution from both prongs.

Note finally that the same result is obtained with the one-pass generalised-k_t axes or with the minimal axes. However, if we were to use exclusive k_t axes, which contrary to the above arguments orders emission's in $z_i \theta_i$, we could have situations where the emission with the largest $z_i \theta_i$ is different from the emission with the largest ρ_i. This inevitably leads to additional complexity.

LL Mass Distribution with a Cut $\tau_{21} < \tau_{\text{cut}}$ Once an expression has been found it is straightforward to understand the structure of the jet mass distribution with a cut $\tau_{21} < \tau_{\text{cut}}$. Since τ_{21} is given by the second "most massive" emission (either from the leading parton or from the emission which dominates the jet mass), imposing a cut on τ_{21} vetoes such emissions, leaving a Sudakov factor corresponding to virtual emissions in that region of phase-space. This is represented on the Lund plane in Fig. 8.1a and one gets

$$\left. \frac{\rho \, d\sigma}{\sigma \, d\rho} \right|_{\tau_{21} < \tau_{\text{cut}}} = \int_0^1 \frac{d\theta_1^2}{\theta_1^2} \frac{dz_1}{z_1} \frac{\alpha_s(z_1 \theta_1) C_i}{\pi} \rho \delta(\rho - \rho_1)$$

$$\times \exp[-R_\tau^{(\text{primary})} - R_\tau^{(\text{secondary})}] \tag{8.8}$$

$$R_\tau^{(\text{primary})} = \int_0^1 \frac{d\theta_2^2}{\theta_2^2} \frac{dz_2}{z_2} \frac{\alpha_s(z_2 \theta_2) C_i}{\pi} \Theta\left(\frac{\rho_2}{\rho} > \tau_{\text{cut}}\right), \tag{8.9}$$

$$R_\tau^{(\text{secondary})} = \int_0^{\theta_1^2} \frac{d\theta_{12}^2}{\theta_{12}^2} \int_0^1 \frac{dz_2}{z_2} \frac{\alpha_s(z_1 z_2 \theta_{12}) C_A}{\pi} \Theta\left(\frac{z_2 \theta_{12}^2}{\theta_1^2} > \tau_{\text{cut}}\right), \tag{8.10}$$

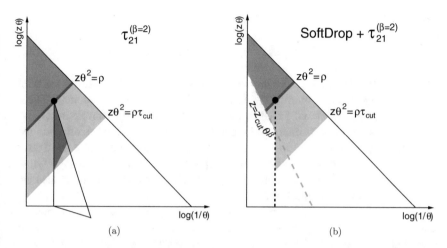

Fig. 8.1 Lund diagram for the LL mass distribution with a cut on the τ_{21} N-subjettiness ratio. The solid red line corresponds to the desired jet mass. Real emissions are vetoed in the shaded light red region because they would yield a larger mass and in the light blue region because they would not pass the cut on τ_{21}. The left plot (**a**) corresponds to the plain jet and the right plot (**b**) to a jet previously groomed with SoftDrop. The left plot (**a**) shows also the plane for secondary (gluon) emissions. An identical secondary plane should also be present on the right plot (**b**) but has been omitted for clarity

where angles are measured in units of the jet radius R and the arguments of the strong couplings are in units of $p_t R$.

The integration in Eq. (8.8) corresponds to the particle which dominates the jet mass, i.e. constrained so that $\rho = \rho_1$. Equation (8.9) is the Sudakov veto on primary emissions. It includes a standard jet-mass Sudakov, $\rho_2 > \rho$, from the fact that emission 1 dominates the mass (the light red region in Fig. 8.1a), as well as an additional Sudakov veto $\rho > \rho_2 > \rho\tau_{\text{cut}}$ coming from the extra constraint on τ_{21}, the light blue region in Fig. 8.1a. Finally, Eq. (8.10) corresponds to the extra Sudakov veto imposing that secondary emissions with $\tau_{21} > \tau_{\text{cut}}$ (cf. Eq. (8.7)) also have to be vetoed. As before, one can obtain the "modified" LL results, including hard collinear splittings, by setting the upper limits of the z integrations to $\exp(B_i)$, which is what we do in practical applications below.

In the fixed-coupling approximation, the integrations can be done analytically, and one obtains

$$\frac{\rho \, d\sigma}{\sigma \, d\rho}\bigg|_{\tau_{21} < \tau_{\text{cut}}} \overset{\text{f.c.}}{=} \frac{\alpha_s C_i}{\pi}(L_\rho + B_i)$$

$$\times \exp\left[-\frac{\alpha_s C_i}{\pi}(L_\rho + L_\tau + B_i)^2 \right.$$

$$\left. -\frac{\alpha_s C_A}{\pi}(L_\tau + B_g)^2 \right], \qquad (8.11)$$

where we have defined

$$L_\rho = \log(1/\rho) \qquad \text{and} \qquad L_\tau = \log(1/\tau_{\text{cut}}). \qquad (8.12)$$

This has to be compared to the jet mass distribution without the cut on τ_{21} which has the same prefactor but only $\frac{\alpha_s C_i}{\pi}(L_\rho + B_i)^2$ in the Sudakov exponent. The cut on τ_{21} brings an additional Sudakov suppression, double-logarithmic in τ_{cut} with contributions from both primary and secondary emissions and, more interestingly, a contribution proportional to $\log(1/\rho)\log(1/\tau_{\text{cut}})$, meaning that with a fixed cut on τ_{21}, the QCD background will be more suppressed when increasing the jet boost, i.e. decreasing ρ. We provide more physical discussions below, once we also have results for the signal and ROC curves.

The calculation of the jet mass with a cut on τ_{21} can also be performed for groomed jets, i.e. one grooms the jet before measuring its mass and τ_{21} on the groomed jet. Here we consider the case of SoftDrop. As discussed in Sect. 6.1, emission 1, which dominates the SoftDrop mass, has to satisfy the SoftDrop condition and the associated Sudakov is given by Eq. (6.10). One small extra complication compared to the case of the SoftDrop jet mass is that one should remember that the SoftDrop de-clustering procedure stops once some hard structure has been found, i.e. once the SoftDrop condition is met. Since the de-clustering procedure uses the Cambridge/Aachen jet algorithm, this means that once the procedure stops, all emissions at smaller angles are kept, whether or not they pass the SoftDrop condition.

In our LL calculation for τ_{21}, it is sufficient to realise that one can consider that the SoftDrop procedure keeps all emissions at angles smaller than θ_1. Thus, the resulting phase-space is depicted in Fig. 8.1b and one gets:

$$\frac{\rho}{\sigma}\frac{d\sigma}{d\rho}\bigg|^{\text{SD}}_{\tau_{21}<\tau_{\text{cut}}} = \int_0^1 \frac{d\theta_1^2}{\theta_1^2} \frac{dz_1}{z_1} \frac{\alpha_s(z_1\theta_1)C_i}{\pi} \rho\delta(\rho - \rho_1)$$

$$\Theta(z_1 > z_{\text{cut}}\theta_1^\beta)e^{-R^{(\text{primary})}_{\tau,\text{SD}} - R^{(\text{secondary})}_\tau} \qquad (8.13)$$

$$R^{(\text{primary})}_{\tau,\text{SD}} = \int_0^1 \frac{d\theta_2^2}{\theta_2^2} \frac{dz_2}{z_2} \frac{\alpha_s(z_2\theta_2)C_i}{\pi}$$

$$\Theta\left(\frac{\rho_2}{\rho} > \tau_{\text{cut}}\right)\Theta(z_2 > z_{\text{cut}}\theta_2^\beta \text{ or } \theta_2 < \theta_1). \qquad (8.14)$$

The Sudakov corresponding to secondary emissions is the same as for the plain jet, since all emissions at angles smaller than θ_1 are kept in the groomed jet. Keeping the running-coupling contributions, one finds the following expressions for the radiators:

$$R_{\tau,\mathrm{SD}}^{(\mathrm{primary})}(\rho, \tau_{\mathrm{cut}}, \theta_1) = R_{\mathrm{SD}}^{(\mathrm{LL})}(\rho\tau_{\mathrm{cut}}) + \delta R_{\tau,\mathrm{SD}}(\rho, \tau_{\mathrm{cut}}, \theta_1) \tag{8.15}$$

$$\delta R_{\tau,\mathrm{SD}}(\rho, \tau_{\mathrm{cut}}, \theta_1) = \frac{C_i}{2\pi\alpha_s\beta_0^2}$$

$$\times \left[W(1 - \lambda_\rho - \lambda_\tau + \lambda_1) + \frac{W(1 - \lambda_c - (1 + \beta)\lambda_1)}{1 + \beta} \right. \tag{8.16}$$

$$\left. - \frac{2 + \beta}{1 + \beta} W\left(1 - \frac{\lambda_c + (1 + \beta)(\lambda_\rho + \lambda_\tau)}{2 + \beta}\right) \right]$$

$$\Theta(\lambda_c + (2 + \beta)\lambda_1 > \lambda_\rho + \lambda_\tau)$$

$$R_\tau^{(\mathrm{secondary})}(\rho, \tau_{\mathrm{cut}}, \theta_1) = \frac{C_i}{2\pi\alpha_s\beta_0^2}$$

$$\times \left[W(1 - \lambda_\rho - \lambda_{B_g} + \lambda_1) + W(1 - \lambda_\rho - \lambda_\tau + \lambda_1) \right. \tag{8.17}$$

$$\left. - 2W(1 - \lambda_c - \frac{\lambda_\tau + \lambda_{B_g}}{2} + \lambda_1) \right]\Theta(\lambda_\tau > \lambda_{B_g}),$$

with λ_ρ and λ_c defined as in Eq. (6.10), $\lambda_\tau = 2\alpha_s\beta_0 \log(1/\tau_{\mathrm{cut}})$ and, $\lambda_1 = 2\alpha_s\beta_0 \log(1/\theta_1)$ and $\lambda_{B_g} = -2\alpha_s\beta_0 B_g$. $\delta R_{\tau,\mathrm{SD}}$ is the additional contribution from $\theta_2 < \theta_1$ and $z_2 < z_{\mathrm{cut}}\theta_2^\eta$. $R_\tau^{(\mathrm{primary})}$ can be easily obtained from $R_{\tau,\mathrm{SD}}^{(\mathrm{primary})}$ by taking the limit $\beta \to \infty$ and it is nothing else than the plain (ungroomed) jet mass Sudakov evaluated at the scale $\rho\tau_{\mathrm{cut}}$. Contrary to the fixed-coupling limit, $\delta R_{\tau,\mathrm{SD}}$ and $R_\tau^{(\mathrm{secondary})}$ explicitly depend on θ_1 and the integration in Eq. (8.13) cannot be performed analytically.

8.1.2 N-Subjettiness Dichroic $\tau_{21}^{(\beta=2)}$ Ratio

The idea behind dichroic observables arises when combining a prong finder and a shape constraint. The identification of two hard prongs in a jet, is usually achieved by applying tools like the mMDT, trimming or pruning to the jet. These algorithms are also active, (and tight) groomers, meaning that they groom away a large fraction of soft and large-angle radiation in the jet. However, the region of phase-space which is groomed away does carry a lot of information about the radiation pattern, which would be potentially exploited by the shape constraint. The idea is therefore to to compute the shape constraint on a larger, less tightly groomed jet, that we call the *large* jet below. For shapes which are expressed as a ratio, like τ_{21}, and for $\beta = 2$, the denominator of the shape is a measure of the jet mass—recall $\tau_1 = \rho$ in the previous section—which is naturally computed on the tight jet found by the prong finder, referred to as the *small* jet in what follows. This hints at the following combination

$$\text{mass constraint: use } \rho_{\text{small}}, \tag{8.18}$$

$$\text{shape constraint: use } \tau_{21}^{(\text{dichroic})} = \frac{\tau_{2,\text{large}}}{\tau_{1,\text{small}}}. \tag{8.19}$$

We will assume that the small jet is obtained using mMDT with the condition $z > x_{\text{cut}}$, and the large jet is either the plain jet or a SoftDrop jet with positive β and a given z_{cut}. We first derive LL analytic results similar to the ones obtained in the previous section for τ_{21} and then come back to the benefits of the dichroic variant.

Approximate $\tau_{21}^{(\text{dichroic})}$ Value at LL The value of $\tau_{21}^{(\text{dichroic})}$ for a given set of emissions in a jet can be readily obtained from the results in the previous section. First, $\tau_{1,\text{small}}$ is equivalent to the small-jet (dimensionless squared) mass: $\tau_{1,\text{small}} = \rho_{\text{small}}$. We will denote by a the emission that sets the mass of the small jet.

For $\tau_{2,\text{large}}$, we need to use Eq. (8.6), i.e. $\tau_{2,\text{large}}$ is dominated by the emission with the second-largest $\rho_i = z_i \theta_i^2$ in the large jet. We will therefore denote by b and c, the emissions with the largest and second-largest ρ_i in the large jet, respectively. With these notations, we get

$$\tau_{21}^{(\text{dichroic})} \approx \frac{\rho_c}{\rho_a} \qquad (\rho_a \text{ largest in small}, \rho_c 2^{\text{nd}} \text{ largest in large}). \tag{8.20}$$

Note that, contrary to the standard τ_{21} ratio, the dichroic ratio can be larger than one. More specifically, three situations can arise: (1) the emission which dominates the mass of the small jet also dominates the one of the large jet, i.e. $\rho_a = \rho_b > \rho_c$, yielding $\tau_{21}^{(\text{dichroic})} < 1$; (2) the emission which dominates the mass of the small jet is the 2nd largest in the large jet, i.e. $\rho_b > \rho_a = \rho_c$ yielding $\tau_{21}^{(\text{dichroic})} = 1$; and (3) there are at least two emissions with a larger ρ_i in the large jet than in the small jet, i.e. $\rho_b > \rho_c > \rho_a$ yielding $\tau_{21}^{(\text{dichroic})} > 1$. It is easy to check that the value of $\tau_{21}^{(\text{dichroic})}$ is always equal or larger than the value of the τ_{21} ratio obtained with approaches frequently used in experimental contexts. This is a desired feature since increasing the value of τ_{21} means rejecting more QCD jets when imposing a cut.[2]

LL Mass Distribution with a Cut $\tau_{21}^{(\text{dichroic})} < \tau_{\text{cut}}$ The calculation of the jet mass distribution with a cut on $\tau_{21}^{(\text{dichroic})}$ has to be separated in the same three possible of mass orderings as before, corresponding to $\tau_{21}^{(\text{dichroic})}$ smaller, equal or larger than 1. The three situations are represented in Fig. 8.2 for the case where the large jet has been groomed with SoftDrop using a positive β.

---

[2]As we will see below, this increase of τ_{21} for QCD jets in the dichroic case comes with no modifications for signal jets.

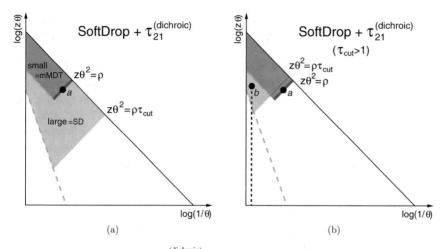

Fig. 8.2 Lund diagrams for a cut $\tau_{21}^{\text{(dichroic)}} < \tau_{\text{cut}}$, assuming that mMDT is used for the small jet and SoftDrop for the large jet. Emissions a and b are the emissions with the largest $z_i \theta_i^2$ in the mMDT and SoftDrop jet respectively. The shaded red region corresponds to the vetoed region from the requirement on the (small) jet mass, and the shaded blue region is the extra Sudakov veto from the constraint on $\tau_{21}^{\text{(dichroic)}}$. Figure (**a**) corresponds to a cut $\tau_{\text{cut}} < 1$ for which emissions a and b are identical. Figure (**b**) corresponds to $\tau_{\text{cut}} > 1$, where one has an emission ρ_b in the large jet such that $\rho_b > \rho$ and one has to veto real emissions with $z\theta^2 > \rho\tau$. In both cases, we omitted a contribution from secondary emissions for readability. It corresponds to a secondary plane originating from emission a (resp. b) in case (**a**) (resp. (**b**)), with a Sudakov veto extending down to $z\theta^2 = \rho\tau_{\text{cut}}$ with z measured with respect to the initial jet

   The case of a cut $\tau_{\text{cut}} < 1$ is the most interesting as it is the situation relevant for phenomenology—the other cases would, as we show below, also kill the signal—and where the effect of adopting a dichroic ratio can be explicitly seen. As for the case of the standard τ_{21}, one as to integrate over the emission a which dominates the small jet mass and veto any additional real emission which would give a value of $\tau_{21}^{\text{(dichroic)}}$ larger than τ_{cut}, i.e. any emission in the large jet with $z\theta^2 > \rho_a \tau_{\text{cut}}$. This gives

$$\frac{\rho}{\sigma} \frac{d\sigma}{d\rho} \bigg|_{\tau_{21} < \tau_{\text{cut}}}^{\text{dichroic}} \stackrel{\tau_{\text{cut}} < 1}{=} \int_0^1 \frac{d\theta_a^2}{\theta_a^2} \frac{dz_a}{z_a} \frac{\alpha_s(z_a\theta_a)C_i}{\pi} \rho\delta(\rho - \rho_a)$$

$$\Theta(z_a > x_{\text{cut}}) e^{-R_{\tau,\text{SD}}^{\text{(primary)}} - R_{\tau}^{\text{(secondary)}}}, \tag{8.21}$$

with $R_{\tau,\text{SD}}^{\text{(primary)}}$ and $R_{\tau}^{\text{(secondary)}}$ again given by (8.14) and (8.10). Compared to Eq. (8.13), one clearly sees that the lower bound of the z_a (z_1 in (8.13)) integration has been increased, corresponding to a reduction of the QCD cross-section in the dichroic case.

   For completeness, we briefly discuss the case $\tau_{\text{cut}} \geq 1$. Situations with zero or one emissions in the large jet with $\rho_b > \rho$ give $\tau_{21}^{\text{(dichroic)}} \leq 1$ and are therefore accepted. For situations with (at least) two emissions $\rho_b > \rho_c > \rho_a$, one only accepts the cases with $\rho_c/\rho < \tau_{\text{cut}}$. Thus, the only situation which has to be vetoed is $\rho_b > \rho_c > \rho\tau_{\text{cut}}$.

This can be reorganised in a slightly more convenient way. First, if there is no emission ρ_b with $\rho_b > \rho\tau_{\text{cut}}$, the veto condition cannot be satisfied, meaning the case always contributes to the cross-section. For cases with at least one emission such that $\rho_b > \rho\tau_{\text{cut}}$, one needs an additional veto on emissions c such that $\rho_b > \rho_c > \rho\tau_{\text{cut}}$. This situation corresponds to Fig. 8.2b. If one assumes that the small jet is obtained using mMDT and the large jet using SoftDrop, and if we denote by R_{out} the radiator corresponding to the region in the large jet but outside the small one (i.e. the shaded blue region in Fig. 8.2), this yields

$$\frac{\rho}{\sigma}\frac{d\sigma}{d\rho}\bigg|_{\tau_{21} < \tau_{\text{cut}}}^{\text{dichroic}} \stackrel{\tau_{\text{cut}} > 1}{=} \int_0^1 \frac{d\theta_a^2}{\theta_a^2}\frac{dz_a}{z_a}\frac{\alpha_s(z_a\theta_a)C_i}{\pi}\rho\delta(\rho - \rho_a)\,\Theta(z_a > x_{\text{cut}}) \qquad (8.22)$$

$$\times \left[e^{-R_{\text{out}}(\rho\tau_{\text{cut}})} + \int_0^1 \frac{d\theta_b^2}{\theta_b^2}\frac{dz_b}{z_b}\frac{\alpha_s(z_b\theta_b)C_i}{\pi}\Theta(\rho_b > \rho\tau_{\text{cut}}) \right.$$

$$\Theta(x_{\text{cut}} > z_b > z_{\text{cut}}\theta_b^\beta)$$

$$\left. \times e^{-R_{\text{out}}(\rho\tau_{\text{cut}},\rho_b,\theta_b)-R_\tau^{(\text{secondary})}(\rho_b,\rho\tau_{\text{cut}}/\rho_b,\theta_b)} \right]$$

In this expression, $R_{\text{out}}(\rho\tau_{\text{cut}})$ is trivially given by $R_{\text{SD}}(\rho\tau_{\text{cut}}) - R_{\text{mMDT}}(\rho\tau_{\text{cut}})$. In the presence of an emission b, one has to be careful that SoftDrop will keep emissions at angles smaller than θ_b, and therefore $R_{\text{out}}(\rho\tau_{\text{cut}},\rho_b,\theta_b) = R_{\tau,\text{SD}}^{(\text{primary})}(\rho_b, \rho\tau_{\text{cut}}/\rho_b, \theta_b) - R_{\text{mMDT}}(\rho\tau_{\text{cut}})$. We note that in (8.22), the integration over z_a can be done explicitly and gives an overall factor $R'_{\text{mMDT}}(\rho)$. Finally, (8.22) does not coincides with (8.21) when $\tau_{\text{cut}} \to 1$. This is simply because situations with a single emission $\rho_b > \rho$ give $\tau_{21}^{(\text{dichroic})} = 1$, yielding a discontinuity at $\tau_{\text{cut}} = 1$, or, equivalently, a contribution to the τ_{21} distribution proportional to $\delta(\tau_{21} - 1)$.

8.1.3 Energy-Correlation Functions $C_2^{(\beta=2)}$ or $D_2^{(\beta=2)}$

The last shape we want to discuss is the energy-correlation-function ratio D_2, or, almost equivalently, C_2 (which differs from D_2 by a factor ρ). As before, we first give an analytic expression, valid in the leading-logarithmic approximation, for the value of D_2 for a given jet. We then compute the mass distribution with a cut $D_2 < D_{\text{cut}}$.

Approximate D_2 Value at LL Consider once again a set of n emissions with momentum fractions z_i and emitted at angles θ_i from the parent parton, and define $\rho_i = z_i\theta_i^2$. We can assume, as before, that the jet mass is dominated by emission 1, i.e. the jet mass is $\rho \approx \rho_1$. From Eq. (5.17) we then have

$$e_3^{(\beta=2)} = \sum_{i<j<k\in\text{jet}} z_i z_j z_k \theta_{ij}^2 \theta_{ik}^2 \theta_{jk}^2 \approx \sum_{i<j} z_i z_j \theta_{ij}^2 \theta_i^2 \theta_j^2 \tag{8.23}$$

$$\approx \sum_{i<j} z_i z_j \max(\theta_i^2, \theta_j^2)\theta_i^2 \theta_j^2 \approx \sum_{i<j} \rho_i \rho_j \max(\theta_i^2, \theta_j^2), \tag{8.24}$$

where, for the second equality we have used the fact that all emissions are soft so we can neglect triplets which do not involve the leading parton, and the third equality comes from the strong angular ordering between emissions, valid at LL.

For pairs i, j which do not include emission 1, we have, assuming $\theta_i \ll \theta_j$, $\rho_i \rho_j \theta_j^2 \ll \rho_1 \rho_j \theta_j^2 < \rho_1 \rho_j \max(\theta_1^2, \theta_j^2)$. These contributions can therefore be neglected and we have

$$e_3^{(\beta=2)} \approx \rho \sum_{i,\theta_i<\theta_1} \rho_i \theta_1^2 + \rho \sum_{i,\theta_i>\theta_1} \rho_i \theta_i^2. \tag{8.25}$$

At LL accuracy, only one emission, that we will denote by "2" will dominate the sum and we have

$$e_3 \approx \rho\rho_2\max(\theta_1^2, \theta_2^2) \quad \Rightarrow \quad D_2 = \frac{e_3}{e_2^3} \approx \frac{\rho_2}{\rho^2}\max(\theta_1^2, \theta_2^2), \tag{8.26}$$

which has an extra factor $\max(\theta_1^2, \theta_2^2)/\rho$ compared to the τ_{21} ratio. Alternatively, we can work with $C_2 = \rho D_2$. Note that D_2 can be larger than 1 and, in this case, the LL approximation refers to $C_2 \ll 1$ which is dominated by $\rho_2 \ll \rho$ and $\theta_{1,2}^2 \ll 1$.

Finally, when imposing a constraint on D_2, we are also sensitive to secondary emissions from 1. A gluon "2" emitted with a momentum fraction z_2 (measured with respect to to z_1) at an angle θ_{12} from 1, will give a (dominant) contribution $z_1^2 z_2 \theta_{12}^2 \theta_1^4$ to e_3 (taking the leading parton, and emissions 1 and 2 as i, j and k). We therefore have

$$D_{2,\text{secondary}} \approx \frac{z_2\theta_{12}^2}{\rho}. \tag{8.27}$$

LL Mass Distribution with a Cut $D_2 < D_{\text{cut}}$ The LL mass distribution with a cut on D_2 proceeds as for τ_{21} above except that now the constraint on the shape will impose a Sudakov vetoing emissions for which $\rho_2\max(\theta_1^2, \theta_2^2) > \rho^2 D_{\text{cut}}$, with $\rho_2 < \rho$.

The corresponding phase-space is represented in Fig. 8.3. We have to consider two regimes. First, for $D_{\text{cut}} < 1$, we have $\rho^2 D/\theta_1^2 < \rho$ for any $\rho < \theta_1^2 < 1$, resulting in the phase-space represented in Fig. 8.3a. Then. for $1 < D_{\text{cut}} < 1/\rho$, one can either have $\rho D_{\text{cut}} < \theta_1^2$ or $\rho D_{\text{cut}} > \theta_1^2$. For the former corresponds one again recovers Fig. 8.3a, but for the latter, only the region $\rho^2 D_{\text{cut}} < \rho_2\theta_2^2 < \rho$ (i.e. $\theta_2^2 > \rho D$), shown in Fig. 8.3b.

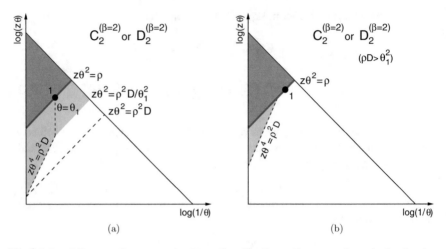

Fig. 8.3 Lund diagrams for a constraint $D_2 < D_{cut}$. For $D_{cut} < 1$, we are always in the situation depicted on figure (**a**), while for $1 < D_{cut} < 1/\rho$, we have either the case of figure (**a**) for $\rho D_{cut} < \theta_1^2$ or the case of figure (**b**) for $\rho D_{cut} > \theta_1^2$. As above, an extra veto for secondary emissions from emission 1 (only in case (**a**)) is not shown for clarity

The mass distribution with a cut on D_2 can be written as

$$\frac{\rho}{\sigma}\frac{d\sigma}{d\rho}\Big|_{D_2<D_{cut}} = \int_0^1 \frac{d\theta_1^2}{\theta_1^2}\frac{dz_1}{z_1}\frac{\alpha_s(z_1\theta_1)C_i}{\pi}\rho\delta(\rho-\rho_1)$$

$$\times \exp[-R_D^{(primary)} - R_D^{(secondary)}] \qquad (8.28)$$

$$R_D^{(primary)} = \int_0^1 \frac{d\theta_2^2}{\theta_2^2}\frac{dz_2}{z_2}\frac{\alpha_s(z_2\theta_2)C_i}{\pi}\Theta\left(\frac{\rho_2}{\rho}\frac{\max(\theta_1^2,\theta_2^2)}{\rho} > D_{cut}\right), \qquad (8.29)$$

$$R_D^{(secondary)} = \int_0^{\theta_1^2} \frac{d\theta_{12}^2}{\theta_{12}^2}\int_0^1 \frac{dz_2}{z_2}\frac{\alpha_s(z_1z_2\theta_{12})C_A}{\pi}\Theta\left(\frac{z_2\theta_{12}^2}{\rho} > D_{cut}\right). \qquad (8.30)$$

For the two cases above, one finds, at LL (including both the mass and shape vetoes)

$$R_D^{(primary)} = \frac{C_i}{2\pi\alpha_s\beta_0^2}$$

$$\times \begin{cases} \frac{1}{3}W(1-2\lambda_\rho-\lambda_D) + \frac{2}{3}W(1-2\lambda_\rho-\lambda_D+\frac{3}{2}\lambda_1) \\ \quad -2W(1-\frac{2\lambda_\rho+\lambda_D-\lambda_1+\lambda_B}{2}) + W(1-\lambda_B) & \text{if } \rho D < \theta_1^2 \\ \frac{1}{3}W(1-2\lambda_\rho-\lambda_D) + \frac{2}{3}W(1-\frac{\lambda_\rho-\lambda_D}{2}) \\ \quad -2W(1-\frac{\lambda_\rho+\lambda_B}{2}) + W(1-\lambda_B) & \text{if } \rho D > \theta_1^2 \end{cases}$$

$$(8.31)$$

$$R_D^{(\text{secondary})} = \frac{C_A}{2\pi\alpha_s\beta_0^2}\left[W\left(1 - 2\lambda_\rho - \lambda_D + \frac{3}{2}\lambda_1\right)\right.$$

$$- 2W\left(1 - \frac{3\lambda_\rho + \lambda_D - 2\lambda_1 + \lambda_{B_g}}{2}\right) + W\left(1 - \lambda_\rho\right)$$

$$\left. - \frac{\lambda_1}{2} + \lambda_{B_g}\right)\right]\Theta(2\lambda_\rho + \lambda_D - \lambda_1 > \lambda_{B_g}) \tag{8.32}$$

where λ_ρ and λ_B are defined as before and we have introduced $\lambda_D = 2\alpha_s\beta_0\log(1/D_{\text{cut}})$ and $\lambda_1 = 2\alpha_s\beta_0\log(1/\theta_1^2)$. $R_D^{(\text{primary})}$ is manifestly continuous at $\rho D = \theta_1^2$.

As for the case of τ_{21}, similar expressions can be obtained with SoftDrop. In this case, the integration over emission 1 in Eq. (8.28) has to be restricted to the region where emission 1 passes the SoftDrop condition. Focusing on the case $\rho < z_{\text{cut}}$, one has, for a given $z_1\theta_1^2 = \rho$, $z_1 > (z_{\text{cut}}^2\rho^\beta)^{\frac{1}{2+\beta}}$ or $\theta_1 < (\rho/z_{\text{cut}})^{\frac{1}{2+\beta}}$. The Sudakov for primary emissions also gets modified by SoftDrop as one only needs to veto emissions for which either $z_2 > z_{\text{cut}}\theta_2^\beta$ or $\theta_2 < \theta_1$. The veto on secondary emissions is unchanged compared to the plain-jet case. After some relatively painful manipulations, one gets

$$R_{D,\text{SD}}^{(\text{primary})} = R_D^{(\text{primary})} - \frac{C_i}{2\pi\alpha_s\beta_0^2}$$

$$\times \left[\frac{1}{3}W(1 - 2\lambda_\rho - \lambda_D) + \frac{W(1 - \lambda_c)}{1+\beta} - \frac{1}{3}W\left(1 - 2\lambda_\rho - \lambda_D + \frac{3}{2}\lambda_1\right)\right.$$

$$\left. - \frac{1}{1+\beta}W\left(1 - \lambda_c - \frac{1+\beta}{2}\lambda_1\right)\right] \tag{8.33}$$

if $\rho D < \theta_1^2$ and $\rho^2 D < z_{\text{cut}}\theta_1^{4+\beta}$,

$$R_{D,\text{SD}}^{(\text{primary})} = R_D^{(\text{primary})} - \frac{C_i}{2\pi\alpha_s\beta_0^2}$$

$$\times \left[\frac{1}{3}W(1 - 2\lambda_\rho - \lambda_D) + \frac{W(1 - \lambda_c)}{1+\beta} - \frac{4+\beta}{3(1+\beta)}\right.$$

$$\left. \times W\left(1 - \frac{(1+\beta)(2\lambda_\rho + \lambda_D) + 3\lambda_c}{4+\beta}\right)\right], \tag{8.34}$$

if either $\rho D < \theta_1^2$ and $\rho^2 D > z_{\text{cut}}\theta_1^{4+\beta}$, or $\rho D > \theta_1^2$ and $z_{\text{cut}}^2\rho^\beta D^{2+\beta} < 1$, and

$$R_{D,\text{SD}}^{(\text{primary})} = R_{\text{SD}}^{(\text{LL})}, \tag{8.35}$$

if $\rho D > \theta_1^2$ and $z_{\text{cut}}^2\rho^\beta D^{2+\beta} > 1$.

The first result corresponds to the situation where one has a contribution similar to $\delta R_\tau^{(SD)}$ in the τ_{21} case, coming from the extra small triangle $z < z_{cut}\theta^\beta$, $\theta < \theta_1$. The existence of this extra region requires $\rho^2 D > z_{cut}\theta_1^{2+\beta}$. The second result with "normal" SoftDrop grooming, covering both kinematic configurations in Fig. 8.3. The third result corresponds to the case of Fig. 8.3b where the shaded blue region is fully outside the region allowed by the SoftDrop condition, in which case, the shape cut has no effects and one recovers a SoftDrop mass Sudakov.

This finishes our calculations for our sample of shapes in the case of QCD jets. Before comparing our results with Monte Carlo simulations, we briefly discuss the case of signal jets so as to be able to discuss the performance when tagging 2-prong boosted objects.

8.1.4 Calculations for Signal Jets

In order to be able to study the performance of two-prong taggers analytically, we also need expressions for signal jets. Generally speaking, signal jets are dominated by the decay of a colourless heavy object of mass m_X into two hard partons. Here, we will assume the decay is in a $q\bar{q}$ pair, which is valid for electroweak bosons W/Z/H and for a series of BSM candidates. If the decay happens at an angle θ_1 (measured in units of the jet radius R) and the quark carries a fraction $1 - z_1$ of the boson's transverse momentum, we have

$$m_X^2 = z_1(1 - z_1)\theta_1^2(p_t R)^2 \quad \text{i.e.} \quad \rho_X = \frac{m_X^2}{p_t^2 R^2} = z_1(1 - z_1)\theta_1^2. \quad (8.36)$$

Furthermore, we will use index 0 (resp. 1) to refer to the quark (resp. antiquark).

The effect of a cut on a jet shape is similar to what we have just discussed for QCD jets: it constrains additional radiation in the jet. The key difference with QCD jets is that now the radiation, is only coming from the $q\bar{q}$ dipole. In the collinear limit sufficient for our purpose here this is equivalent to having two secondary-like Lund planes associated with the quark and antiquark respectively, as depicted on Fig. 8.4.

Calculation of the Shape Value The calculation for a given shape proceeds as before by first computing an expression for the shape value. Say that emission 2, emitted at an angle θ_{02} from the quark (or θ_{12} from the antiquark) and carrying a fraction x_2 of the jet's transverse momentum, dominates the shape value (at LL). For the N-subjettiness τ_{21} ratio, the two axes will align with the quark and antiquark and we find

$$\tau_2 = x_2\min(\theta_{02}^2, \theta_{12}^2) \quad \Rightarrow \quad \tau_{21} \approx \frac{x_2\min(\theta_{02}^2, \theta_{12}^2)}{\rho}. \quad (8.37)$$

Fig. 8.4 Lund plane for signal jets. The two solid dots correspond to the initial $a\bar{a}$ splitting which satisfies $z_1(1-z_1)\theta_1^2 = \rho$. A Lund plane originates from each of the two quarks and the shape constraints impose Sudakov vetos (represented as the shaded areas) in each of them

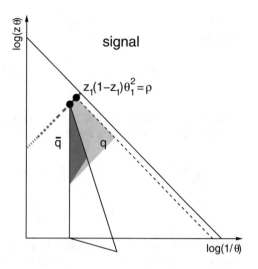

This expression is also valid for the dichroic ratio (just like the contribution from secondary emission for QCD jets). For ECFs, we get

$$e_3 = z_1(1-z_1)x_2\theta_{01}^2\theta_{02}^2\theta_{12}^2$$

$$\approx \rho\theta_{01}^2 x_2\min(\theta_{02}^2, \theta_{12}^2) \quad \Rightarrow \quad D_2 \approx \frac{\theta_{01}^2}{\rho}\frac{x_2\min(\theta_{02}^2, \theta_{12}^2)}{\rho}. \tag{8.38}$$

Signal Efficiency In the case of signal jets with a fixed jet mass, one should compute directly the signal efficiency, i.e. the fraction of signal jets that are accepted by the tagger and the cut on the shape. Assuming again that the two hard prongs are identified using SoftDrop as a prong finder, one can write

$$\epsilon_S(v < v_{\text{cut}}) = \int_0^1 dz_1 P_X(z_1)\Theta(\min(z_1, 1-z_1) > z_{\text{cut}}\theta_{01}^\beta)$$

$$\times e^{-R_v^{(q)}(v_{\text{cut}};z_1)-R_v^{(\bar{q})}(v_{\text{cut}};z_1)}, \tag{8.39}$$

where $P_X(z_1)$ is the probability density for the quark to carry a fraction $1-z_1$ of the boson's transverse momentum (for simplicity, we will assume $P_X(z) = 1$ in what follows), θ_{01} is constrained by Eq. (8.36), and the veto on radiations in the quark and antiquark prongs takes the form of a Sudakov suppression, with the two related by a $z_1 \leftrightarrow 1-z_1$ symmetry $R_v^{(\bar{q})}(v; z_1) = R_v^{(q)}(v; 1-z_1)$. As already discussed in Sect. 6.4, an important aspect of signal jets is that P_X is finite when z_1 or $1-z_1$ goes to 0.

Note that from the above signal efficiency, one can recover the differential distribution of the shape value using

$$\frac{v\, d\sigma}{\sigma\, dv}\bigg|_{\text{signal}} = \frac{1}{\epsilon_S(\text{no } v \text{ cut})} \frac{d\epsilon_S(v < v_{\text{cut}})}{d\log(v_{\text{cut}})}\bigg|_{v_{\text{cut}}=v}. \tag{8.40}$$

The Sudakov exponents can be computed explicitly for the τ_{21} ratio and D_2. For τ_{21} ("standard" or dichroic), we find, using $x_2 = (1 - z_1)z_2$

$$R_v^{(\bar{q})}(v; z_1) = \int_0^{\theta_{01}^2} \frac{d\theta_{02}^2}{\theta_{02}^2} \int_0^1 \frac{dz_2}{z_2} \frac{\alpha_s(x_2\theta_{02})C_F}{\pi}$$

$$\Theta((1 - z_1)z_2\theta_{02}^2 > \rho\tau_{\text{cut}})\Theta((1 - z_1)^2 z_2\theta_{02}^2 < \rho), \tag{8.41}$$

where the last condition of the first line imposes that emission 2 does not dominate the mass.[3] At leading logarithmic accuracy, including as well hard-collinear splittings by imposing $z_2 < \exp(B_q)$ as before, one gets (with $\log(1/\tau) + B_q > 0$)

$$R_\tau^{(\bar{q})}(v; z_1) \overset{\text{LL}}{=} \frac{C_F}{2\pi\alpha_s\beta_0^2}\left\{\left[W\left(1 - \frac{\lambda_\rho - \lambda_z + \lambda_-}{2} - \lambda_B\right)\right.\right.$$

$$- 2W\left(1 - \frac{\lambda_\rho + \lambda_- + \lambda_\tau + \lambda_B}{2}\right)$$

$$\left. + W\left(1 - \frac{\lambda_\rho + \lambda_z + \lambda_-}{2} - \lambda_\tau\right)\right]$$

$$- \left[W\left(1 - \frac{\lambda_\rho - \lambda_z + \lambda_-}{2} - \lambda_B\right)\right. \tag{8.42}$$

$$- 2W\left(1 - \frac{\lambda_\rho + \lambda_B}{2}\right)$$

$$\left.\left. + W\left(1 - \frac{\lambda_\rho + \lambda_z - \lambda_-}{2}\right)\right]\Theta(\lambda_z - \lambda_- > \lambda_B)\right\},$$

where $\lambda_z = 2\alpha_s\beta_0 \log(1/z_1)$ and $\lambda_- = 2\alpha_s\beta_0 \log(1/(1 - z_1))$.

---

[3]This is mostly an artefact of our approximations. In the case of signal jets with $z_1 \ll 1$, this is equivalent to saying that the effect of the shape corresponds to the shaded blue region in Fig. 8.1 which extends up to $z\theta^2 = \rho$, with the region above corresponding to the structure which gives the mass. In practice, this condition is valid up to finite squared logarithms of $1 - z_1$ when $1 - z_1 > z_1$, i.e. well beyond our current accuracy.

For D_2 we find similarly (with $\log(1/\tau) + B_q > \log(z_1^2(1 - z_1))$)

$$R_D^{(\bar{q})}(v; z_1) = \int_0^{\theta_{01}^2} \frac{d\theta_{02}^2}{\theta_{02}^2} \int_0^1 \frac{dz_2}{z_2} \frac{\alpha_s(x_2\theta_{02})C_F}{\pi}$$

$$\Theta\left(\frac{z_2\theta_{02}^2}{z_1} > \rho D\right)\Theta((1 - z_1)^2 z_2\theta_{02}^2 < \rho)$$

$$\stackrel{\text{LL}}{=} \frac{C_F}{2\pi\alpha_s\beta_0^2}\left\{\left[W\left(1 - \frac{\lambda_\rho - \lambda_z + \lambda_-}{2} - \lambda_B\right)\right.\right.$$

$$- 2W\left(1 - \frac{\lambda_\rho + \lambda_z + 2\lambda_- + \lambda_D + \lambda_B}{2}\right)$$

$$\left.+ W\left(1 - \frac{\lambda_\rho + 3\lambda_z + 3\lambda_-}{2} - \lambda_D\right)\right]$$

$$- \left[W\left(1 - \frac{\lambda_\rho - \lambda_z + \lambda_-}{2} - \lambda_B\right)\right.$$

$$- 2W\left(1 - \frac{\lambda_\rho + \lambda_B}{2}\right)$$

$$\left.\left.+ W\left(1 - \frac{\lambda_\rho + \lambda_z - \lambda_-}{2}\right)\right] \qquad (8.43)$$

$$\left.\Theta(\lambda_z - \lambda_- > \lambda_B)\right\},$$

These expressions will be compared to Monte Carlo simulations in the next section where we also discuss key phenomenological observations.

8.2 Comparison to Monte Carlo Simulations

In this section, we compare our analytic results to Monte Carlo simulations obtained with Pythia. For all the Monte Carlo simulations in this chapter, we have relied on the samples used in the "two-prong tagger study" performed in the context of the Les Houches Physics at TeV Colliders workshop in 2017 (Section III.2 of Ref. [122]). Background jets are obtained from a dijet sample while signal jets are obtained from a WW event sample.

In order to streamline our discussion, we focus on a selection of five working points:

- $\tau_{21}^{(\text{SD})}$: SoftDrop jet mass with a cut on $\tau_{21}^{(\beta=2)}$ computed on the SoftDrop jet;
- $\tau_{21}^{(\text{mMDT})}$: mMDT jet mass with a cut on $\tau_{21}^{(\beta=2)}$ computed on the mMDT jet;
- $\tau_{21}^{(\text{dichroic})}$: mMDT jet mass with a cut on $\tau_{21}^{(\text{dichroic})} = \tau_2^{(\text{SD})}/\tau_1^{(\text{mMDT})}$;

- $D_2^{(SD)}$: SoftDrop jet mass with a cut on $D_2^{(\beta=2)}$ computed on the SoftDrop jet;
- $D_2^{(mMDT)}$: mMDT jet mass with a cut on $D_2^{(\beta=2)}$ computed on the mMDT jet.

Note that above selection of working points never includes the plain jet. Although using ungroomed jets can show good tagging performances (as expected from the discussion below), they usually have poor resilience against non-perturbative effects (see next section) and are therefore less relevant for a comparison to analytic calculations.

We first focus on the shape distributions, shown for QCD and signal (W) jets in Fig. 8.5. Globally speaking, we see that the main features observed in the Monte

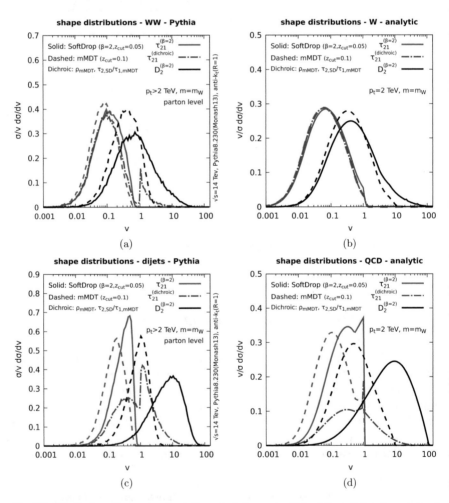

Fig. 8.5 Distributions for our representative set of shapes as obtained from Pythia (**a** and **c**) and from the analytic calculations of Sect. 8.1 (**b** and **d**). The top row corresponds to signal (WW) jets while the bottom row shows results for background (QCD) jets

Carlos simulations are well reproduced by our simple analytic calculations, although the former exhibit distributions that are generally more peaked than the ones obtained with the analytics. We observe that the signal distribution is, to a large extend, independent of the level of grooming (SoftDrop or mMDT). Analytically, this comes from the fact that the grooming procedure stops at the angle θ_{01} of the $W \to q\bar{q}$ decay, keeping the full radiation inside the two prongs unaffected by the groomer. The small differences seen in the Pythia simulations are likely due to radiation outside the $q\bar{q}$ prongs and to initial-state radiation which is less efficiently groomed by SoftDrop (with $\beta = 2$) than by mMDT, shifting the former to slightly larger values than the latter. In the case of D_2, the differences between the SoftDrop and mMDT results also involve the fact that the D_2 Sudakov has a stronger dependence on the p_t sharing between the quark and antiquark than τ_{21}. A specific case of this independence of signal distributions to grooming is that the distribution for the dichroic τ_{21} ratio is very close to the "standard" ones, again with little differences seen e.g. by the presence of a small peak at $\tau_{21} > 1$.

Turning to QCD jets, the situation is clearly different: distributions shift to smaller values when applying a tighter grooming i.e. when going from SoftDrop to mMDT. This shift is reasonably well reproduced in the analytic calculation and it is due to the fact that jet shapes are sensitive to radiations at large angles—larger than the angle of the two-prong decay dominating the jet mass—which is present in QCD jets, but largely absent in W jets. *This has a very important consequence: one expects the tagging efficiency to increase for lighter grooming on the jet shape* as the signal is largely unmodified and the background peak is kept at large values of the shape. In this context, the case of the dichroic N-subjettiness ratio is also interesting: the dichroic distribution (mixing mMDT and SoftDrop information) has larger values than both the corresponding SoftDrop and mMDT distributions. In other words, at small τ_{21}, relevant for tagging purposes, the dichroic distribution is lower than the SoftDrop and mMDT ones. From an analytic viewpoint, one expects the dichroic distribution to be smaller than the SoftDrop distribution because, for the same Sudakov suppression, it imposes a tighter condition on the emission that gives the mass, and smaller than the mMDT distribution because keeping more radiation at larger angles increases the Sudakov suppression (cf. Figs. 8.1b and 8.2a). *This is our second important observation: one expects dichroic ratios to give a performance improvement.*

One last comment about Fig. 8.5 is the presence of peaks for $\tau_{21}^{(\text{dichroic})} \gtrsim 1$ in the Pythia simulation and spikes at $\tau_{21}^{(\text{dichroic})} = 1$ in our analytic calculation. As discussed in the analytic calculation of Sect. 8.1.2, the cumulative $\tau_{21}^{(\text{dichroic})}$ distribution is discontinuous at $\tau_{21}^{(\text{dichroic})} = 1$ and this directly gives a $\delta(\tau_{21}^{(\text{dichroic})} - 1)$ contribution to Fig. 8.5d.[4] Once we go beyond leading logarithmic approximation— for example, following the technique introduced in [157]—this is replaced by a Sudakov peak corresponding to what is seen in the Pythia simulations from

---

[4]For readability, the peak has been scaled down on the plot.

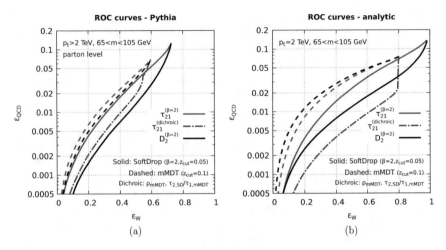

Fig. 8.6 ROC curves corresponding to our representative set of shapes as obtained from Pythia (**a**) and from the analytic calculations (**b**)

Fig. 8.5c. We also note a kink in the τ_{21} and $\tau_{21}^{(\text{dichroic})}$ distributions around 0.5. This corresponds to the point below which secondary emissions start to contribute, namely $\log(\tau_{21}) = B_g$. Since this is in a region where our approximation $\tau_{21} \ll 1$ is not clearly satisfied, subleading corrections play an important role.

We now turn to a direct analysis of the tagging performance of our tools with the ROC curves shown on Fig. 8.6. Note that the tagging efficiencies include both the effect of the requirement on the jet mass and of the cut on the jet shape. For signal jets, we have assumed that the jet mass is exactly the W mass if the jet passes the z_{cut} (or SoftDrop) condition on the W $\rightarrow q\bar{q}$ decay. The two important features highlighted above are indeed seen here: decreasing the level of grooming results in an increased tagging performance, as does using dichroic ratios. In the first case, note that the situation is more delicate at large signal efficiency (close to the endpoint of the ROC curves corresponding to no constraint on the jet shape) since one also has to include the effect of the groomer on constraining the jet mass. Note also that our analytic calculation generally overestimates the signal efficiency, which is likely due to various oversimplifications mentioned earlier.

The other important observation (our third) is that a constraint on D_2 outperforms a constraint on τ_{21}.[5] Although there is only a small gain (that the simple analytic calculation fails to capture) with tight (mMDT) grooming, there is a clear gain in using D_2 when using a looser grooming (SoftDrop) i.e. when opening to larger angles. This is seen in both the analytic calculation and Monte Carlo simulations. This feature can be explained from our analytic approach. Based on

---

[5]We refer here to the standard definition of τ_{21}. A proper assessment of the dichroic ratio would also require using a dichroic version of D_2 which is done in the next section.

Fig. 8.4, fixing the signal efficiency (say for a given z_1 or, equivalently, θ_1) is equivalent to selecting how much of the radiation is vetoed, i.e. fixing the lower end of the shaded blue region. This, in turns, determines the behaviour at small angles ($\theta < \theta_1$) in the case of background jets. The remaining differences between τ_{21} and D_2 therefore comes from radiation at angles larger than θ_1. For the latter, D_2 clearly imposes a stronger constraint (related to its $z\theta^4$ behaviour) than τ_{21} (with a lighter $z\theta^2$) behaviour, cf. Figs. 8.1a and 8.3a.

To conclude this section, we want to make a final comment on two other observations emerging from the analytic results. First, in the case of groomers (used here to find the two prongs dominating the jet mass) we had a strong Sudakov suppression of QCD jets for a relatively mild (typically $1 - 2z_{\text{cut}}$) suppression for signal jets (cf. Chap. 6). In contrast, imposing a cut on a jet shape yields a Sudakov suppression for both the background and the signal. This means that if we want to work at a reasonable signal efficiency, the cut on the shape should not be taken too small. Our analytic calculations, strictly valid in the limit $v_{\text{cut}} \ll 1$ are therefore only valid for qualitative discussions and a more precise treatment is required for phenomenological predictions. We refer to Refs. [157, 188] for practical examples.

Our last remark is also *our last important point: for a fixed mass and cut on the jet shape, the signal efficiency will remain mostly independent of the jet p_t but background jets will be increasingly suppressed for larger p_t.* Analytically, the associated Sudakov suppression in the signal is independent of ρ. For background jets, the Sudakov exponent increases with the boost like $\log(1/\rho)$ (cf. Eq. (8.11), with a similar result for D_2). Note that this dependence on ρ of the background efficiency is not always desirable. In particular, it might complicate the experimental estimation of the background, thus negatively impacting searches for bumps on top of it. An alternative strategy consists of designing a "decorrelated tagger" [155] (see description in Sect. 5.5), e.g. built from ρ and τ_{21}, yielding a flat background, hence facilitating searches.

8.3 Performance and Robustness

The last set of studies we want to perform in this chapter is along the lines of our quality criteria introduced in Sect. 5.2, namely looking at two-prong taggers both in terms of their performance and in terms of their resilience against non-perturbative effects. An extensive study has been pursued in the context of the Les Houches Physics at TeV Colliders workshop in 2017 (LH-2017), where a comparison of a wide range of modern two-prong taggers was performed. Here, we want to focus on a subset of these results, highlighting the main features and arguments one should keep in mind when designing a two-prong tagger and assessing its performance. We refer to Section III.2 of Ref. [122] for additional details and results.

The study is done at three different values of p_t (500 GeV, 1 and 2 TeV) and here we focus on jets reconstructed with the anti-k_t algorithm with $R = 1$ (the LH-2017 study also includes $R = 0.8$). Crucially, we are going to discuss in detail the resilience with respect to non-perturbative effects, including both hadronisation and the Underlying Event. We refer to the extensive study for a separate analysis of hadronisation and the UE, as well as for a study of resilience against detector effects and pileup.

To make things concrete, we consider a wide set of two-prong taggers which can be put under the form

$$
m_{\min} < m < m_{max} \qquad \text{shape } v = \frac{\text{3-particle observable}}{\text{2-particle observable}} < v_{\text{cut}}, \qquad (8.44)
$$

where the mass, the two-particle observable and the three-particle observable can potentially be computed with different levels of grooming. We will focus on four levels of grooming

- *plain (p)*: no grooming,
- *loose (ℓ)*: SoftDrop with $\beta = 2$ and $z_{\text{cut}} = 0.05$,
- *tight (t)*: mMDT with $z_{\text{cut}} = 0.1$,
- *trim*: trimming with k_t subjets using $R_{\text{trim}} = 0.2$ and $f_{\text{trim}} = 0.05$,

and four different shapes: the τ_{21} N-subjettiness ratio and the D_2, N_2 and M_2 ECF ratios either with $\beta = 1$ or $\beta = 2$. A generic tagger can then be put under the form

$$
v\left[m \otimes \frac{n}{d} \right], \qquad (8.45)
$$

where v is one of our three shapes, m is the level of grooming used to compute the jet mass and n and d are the levels of grooming used respectively for the numerator and denominator of the shape. We consider the combinations listed in Table 8.1.

In order to study the tagging quality, we impose the reconstructed mass to be between 65 and 105 GeV and we vary the cut on the jet shape. We select a working point so that the signal efficiency (at truth, i.e. hadron+UE level) is 0.4, which

Table 8.1 List of the different tagging strategies considered with the corresponding level of grooming for the mass, and numerator and denominator of the shape variable

Notation: $m \otimes n/d$	m (mass)	n (numerator)	d (denominator)
$p \otimes p/p$	Plain	Plain	Plain
$\ell \otimes p/p$	Loose	Plain	Plain
$\ell \otimes p/\ell$	Loose	Plain	Loose
$\ell \otimes \ell/\ell$	Loose	Loose	Loose
$t \otimes p/p$	Tight	Plain	Plain
$t \otimes \ell/\ell$	Tight	Loose	Loose
$t \otimes p/t$	Tight	Plain	Tight
$t \otimes \ell/t$	Tight	Loose	Tight
$t \otimes t/t$	Tight	Tight	Tight
trim	Trim	Trim	Trim

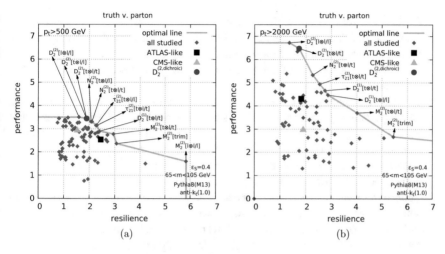

Fig. 8.7 Summary of the performance (significance) v. robustness (resilience) of a set of two-prong taggers based on the combination of a prong finder and a shape cut. The two plots correspond to two different p_t cuts: (**a**) $p_t > 500$ GeV, (**b**) $p_t > 2$ TeV

fixes the cut on the jet shape. For that cut, we can compute both the signal and background efficiencies at parton level and at hadron+UE level, which allows us to compute the tagging performance and robustness using the significance $\epsilon_S/\sqrt{\epsilon_B}$ and resilience ζ introduced in Sect. 5.2. The resulting tagging qualities are summarised in Fig. 8.7 for the two extreme p_t values. Each point on the plot represents a different tagger. The "ATLAS-like" tagger, i.e. trimmed mass with $D_2^{(1)}$ computed on the trimmed jet), and "CMS-like" tagger, i.e. mMDT mass with $N_2^{(1)}$ computed on the mMDT jet, correspond to the working points defined in Sect. 5.5. The $D_2^{(2,\text{dichroic})}$ tagger corresponds to a working point which appears to show a large performance without sacrificing too much resilience. This tagger features $t \otimes \ell/t$ dichroic D_2 variable (with angular exponent $\beta = 2$) with the mass computed on the tight jet, the shape numerator $e_3/(e_e^2)$ computed on the loose jet, and the shape denominator e_2 computed on the tight jet. The plot also shows the line corresponding to the envelope which maximises resilience for a given performance (and vice versa).

There are already a few interesting observations we can draw from Fig. 8.7.

- As p_t increases, the discriminating power increases as well. This can be explained by the fact that when p_t increases, the phase-space for radiation becomes larger, providing more information that can be exploited by the taggers;
- The main observations from the previous section still largely hold: dichroic variants and variants based on D_2 give the best performance. One possible exception is the case of $D_2^{(2)}[\ell \otimes \ell/\ell]$ (i.e. both the mass and D_2 computed on the loose (SoftDrop) jet), which shows a slightly larger performance than our

$D_2^{(2,\text{dichroic})}$ working point, albeit with a smaller resilience.[6] One aspect which is to keep in mind here is that using a looser grooming to measure the jet mass could have the benefit of avoiding the $1 - 2z_{\text{cut}}$ signal efficiency factor before any shape cut is applied, of course probably at the expense of more distortion of the W peak.

- Generically speaking, *there is a trade-off between resilience and performance.* This is particularly striking if one looks along the optimal line. This is an essential feature to keep in mind when designing boosted-object taggers: keeping more radiation in the jet (by using a looser groomer) or putting tighter constraints on soft radiation at larger angles typically leads to more efficient taggers but at the same time yields more sensitivity to the regions where hadronisation and the Underlying Event have a larger impact, hence reducing resilience. This is seen both in terms of the shape, when going from M_2 to τ_{21} and N_2 and then to D_2, and in terms of grooming, when going from tight to loose jets.
- Apart from a few exceptions at relatively lower significance and high resilience, the taggers on the optimal are dominated by shapes with angular exponent $\beta = 2$ rather than the current default at the LHC which is $\beta = 1$.

In order to gain a little more insight than what is presented in the summary plot from Fig. 8.7, we have extracted a few representative cases in Fig. 8.8, where each plot shows different shapes for a fixed grooming strategy and Fig. 8.9 where each plot shows different grooming strategies for a fixed shape. All of the key points made above are visible on these plots. We highlight here a few additional specific examples.

On Fig. 8.8, one sees that the performance of the taggers increases with p_t, with D_2 having the best performance, followed by τ_{21} and N_2 which show a similar pattern, and M_2 which shows a (much) lower performance. With tight grooming, Fig. 8.8a, the phase-space available for radiation constraint is limited and the differences between the shapes are not large. Conversely, when opening more phase-space, e.g. Fig. 8.8c and d, the differences between shapes becomes more visible. The trade-off between performance and resilience is visible in each plot, with the exception of $D_2^{(2)}[\ell \otimes \ell / \ell]$ in Fig. 8.8d. We also see that shapes with angular exponent $\beta = 2$ show a better performance than their $\beta = 1$ counterparts. In terms of resilience which can be either smaller (e.g. $D_2^{(2)}[\ell \otimes \ell / \ell]$), similar, or larger (e.g. $N_2^{(2)}[t \otimes \ell]/t$). We note that for plain jets, we would expect $\beta = 1$ shapes, typically behaving like a k_t scale, to be more resilient than shapes with $\beta = 2$, behaving like a mass scale instead, since they can maximise the available perturbative phase-space before hitting the hadronisation scale (which corresponds to a soft k_t scale). And a similar argument hold for the Underlying Event. Conversely, from a perturbative QCD point of view, $\beta = 2$ has often be shown (see e.g. [121, 143]) to have a

---

[6]If we were seeking absolute performance without any care for resilience, this suggests that even looser groomers, possibly combined with a dichroic approach, could yield an even greater performance.

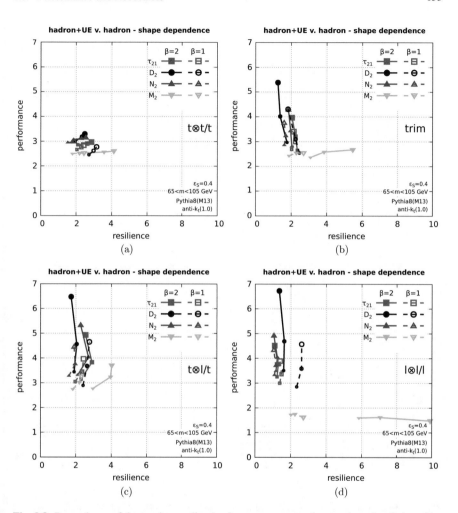

Fig. 8.8 Dependence of the tagging quality (performance versus robustness) on the choice of jet shape. Results are shown for different grooming strategies indicated on each plot. Each curve has three points with increasing symbol size corresponding to $p_t = 500$ GeV, 1 and 2 TeV. Each panel corresponds to a different grooming level as indicated on the plots

larger discriminating power. A natural expectation is therefore that once jets are groomed, non-perturbative these effects are expected to be reduced, giving more prominence to the perturbative QCD tendency to favour $\beta = 2$. Turning finally to Fig. 8.9, we clearly see for all four shapes, that using a looser groomer for the shape (either via the "all-loose" $\ell \otimes \ell / \ell$ or the "dichroic" $t \otimes \ell / t$ combination) comes with large gains in terms of performance. However, using the plain jet typically shows bad performance, an effect which can be attributed to an enhanced sensitivity to the Underlying Event. Comparing the "all-loose" and the "dichroic" variants, we

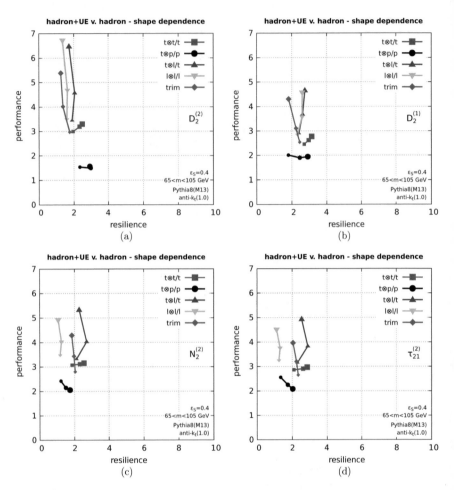

Fig. 8.9 Dependence of the tagging quality (performance versus robustness) on the choice of grooming strategy. Results are shown for a representative set of jet shapes. Each curve has three points with increasing symbol size corresponding to $p_t = 500$ GeV, 1 and 2 TeV. Each panel corresponds to a different choice of shape as indicated on the plots

see that they show a similar performance, with the dichroic variant having a larger resilience.

To conclude, we stress once again that, in order to get a complete picture, the above discussion about performance versus resilience should be supplemented by a study of the resilience against detector effects and pileup. Even though we will not do this study here, one can at least make the educated guess that pileup effects would be reduced by using a tighter grooming.

Chapter 9
Curiosities: Sudakov Safety

In Chap. 5 we have introduced the modified Mass Drop Tagger/SoftDrop and in Chap. 6 we have discussed at length the analytic properties of the jet mass distribution after mMDT or SoftDrop. Furthermore, we have just analysed some aspects of applying this grooming technique to jet shapes used for quark/gluon and W-boson discrimination. However, if we go back to its original definition, we notice that the SoftDrop condition Eq. (5.3) does not involve directly the jet mass or any jet shape, but rather the distance between two prongs in the azimuth-rapidity plane R_{ij} and the momentum fraction $z = \frac{\min(p_{t,i}, p_{t,j})}{p_{t,i} + p_{t,j}}$. It is quite natural to ask ourselves if we can apply the calculation techniques described for jet masses and jet shapes to better characterise the distributions of these two quantities. To be precise, let us define the two observables θ_g and z_g as follows. We start with a jet which has been re-clustered with Cambridge/Aachen and we apply SoftDrop. When we find the first declustering with subjets j_1 and j_2 that passes the SoftDrop condition Eq. (5.3), we define the groomed radius and the groomed momentum fraction as

$$\theta_g = \frac{R_{12}}{R}, \tag{9.1}$$

$$z_g = \frac{\min(p_{t,1}, p_{t,2})}{p_{t,1} + p_{t,2}}, \tag{9.2}$$

where R is the original jet radius. We note that these variables are interesting for a number of reasons. The groomed jet radius is of interest because the groomed jet area is of the order of $\pi \theta_g^2$. Thus, θ_g serves as a proxy for the sensitivity of the groomed jet to possible contamination from pileup [64, 191]. Furthermore, as we shall shortly see, z_g provides us with an almost unique perturbative access to one of the most fundamental building blocks of QCD, namely the Altarelli-Parisi splitting function [51, 192].

© Springer Nature Switzerland AG 2019
S. Marzani et al., *Looking Inside Jets*, Lecture Notes in Physics 958,
https://doi.org/10.1007/978-3-030-15709-8_9

This last observation has drawn the interest of the scientific community in particular with the study of heavy-ion collisions. In particular, an observable such as z_g provides information about how perturbative QCD evolution is modified by the interaction between the high-energy jet and the quark-gluon plasma, thus providing a new probe of the latter. Different experiments have now measured z_g distribution. For instance the STAR collaboration at the Relativistic Heavy Ion Collider of the U.S. Brookhaven National Laboratory performed this measurement using gold-gold collisions [14]. Furthermore, the CMS experiment and the ALICE experiments studied this variable, at the Large Hadron Collider, in lead-lead heavy-ion collisions [193, 194]. We will describe some of these measurements in more detail in Chap. 10. In parallel, this line of research triggered noticeable interest in the theoretical nuclear physics and heavy-ion communities, e.g. [195–203].

In this chapter, we focus on a baseline description of the θ_g and z_g observables in proton-proton collision, leaving aside the extra complications due to interactions of jets with the quark-gluon plasma in the heavy-ion case. In this context, we anticipate that while we will be able to apply the standard techniques presented so far in this book in order to obtain a perturbative prediction for the θ_g distribution for, the situation will be very different for z_g, where interesting features emerge.

9.1 The Groomed Jet Radius Distribution θ_g

We start by calculating the cumulative distribution for the groomed jet radius. In doing so, we are going to exploit the techniques developed in the previous chapters. In particular, we begin by drawing the Lund plane for the observables at hand. We do this in Fig. 9.1, where we distinguish three cases according to the sign of the SoftDrop angular exponent β. From left to right we have $\beta < 0$, $\beta = 0$ and $\beta > 0$. We remind the reader that SoftDrop with $\beta = 0$ corresponds to mMDT.

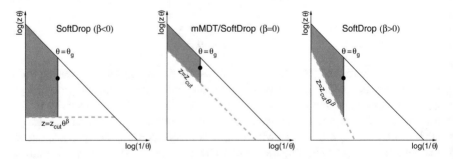

Fig. 9.1 Lund diagrams for the θ_g distribution for three representative values of the SD angular exponent β. From left to right we have $\beta < 0$, $\beta = 0$ (mMDT) and $\beta > 0$. The dashed green line represents the edge of SD region, the solid red line corresponds to emissions yielding the requested groomed jet radius and the shaded red area is the vetoed area associated with the Sudakov suppression. We note that the latter is finite in all three cases, as it should be for an IRC observable

The dashed green line represents the edge of phase-space region where emissions pass the SoftDrop condition, while the solid red line corresponds to emissions yielding the requested groomed jet radius. Finally, the shaded red area is the region we have to veto in order not to exceed the requested groomed radius. With these considerations and the expertise gained from the previous chapters, we can almost immediately arrive at an all-order cumulative distribution, which resums leading logarithms and next-to-leading ones but limited to the collinear sector. We have

$$\Sigma_{SD}(\theta_g) = \exp\left[-\int_{\theta_g}^1 \frac{d\theta}{\theta} \int_0^1 dz\, P_i(z) \frac{\alpha_s(z\theta p_t R)}{\pi} \Theta\left(z > z_{cut}\theta^\beta\right)\right]$$

$$\equiv \exp\left[-R(\theta_g)\right], \tag{9.3}$$

where the integral in the exponent again corresponds to vetoed emissions and $i = q, g$ depending on the jet flavour. We note that the integrals in Eq. (9.3) are finite (modulo the question of the Landau pole) for all values of β. This is the case because the integral in the exponent arises after adding together real and virtual contributions and therefore its finiteness is guaranteed by the IRC safety of the observable. The integrals in Eq. (9.3) can be easily evaluated to leading-logarithmic accuracy, leading to the following radiator[1]

$$R(\theta_g) = \frac{C_i}{2\pi\alpha_s\beta_0^2}\left[W(1-\lambda_B) - W(1-\lambda_g-\lambda_B) - \frac{W(1-\lambda_c)}{1+\beta}\right.$$

$$\left. + \frac{W(1-\lambda_c-(1+\beta)\lambda_g)}{1+\beta}\right], \tag{9.4}$$

where $\lambda_g = 2\alpha_s\beta_0 \log\left(\frac{1}{\theta_g}\right)$, $\lambda_c = 2\alpha_s\beta_0 \log\left(\frac{1}{z_{cut}}\right)$ and $\lambda_B = -2\alpha_s\beta_0 B_i$ as before.

For $\beta < 0$, this distribution has an endpoint at $\theta_g^{(min)} = z_{cut}^{-1/\beta}$ (modulo corrections from hard-collinear splittings). Correspondingly, there is a finite probability, $\exp[-R(\theta_g^{(min)})]$, that the SoftDrop de-clustering procedure does not find a two-prong structure passing the SoftDrop condition, in which case we set $\theta_g = 0$.

The theoretical calculation is compared to the Monte Carlo prediction, at parton level, in Fig. 9.2, showing that it captures the main features of the distribution. In particular, we notice that the θ_g distribution has an endpoint for negative values of β, related to the finiteness of the available phase-space. Furthermore, as β decreases, the distribution is shifted towards smaller angles.[2] Since the groomed jet area is proportional to $\pi\theta_g^2$, this agrees with the expectation that smaller β corresponds to

---

[1] Note that we have used the same approach as for the rest of this book and included it in the double-logarithmic terms. In this specific case, this is less relevant as the endpoint of the distribution does not depend on it, so we could have left it explicitly as a separate correction.

[2] Here, the case of negative β can be seen as shifting a whole part of the distribution to $\theta_g = 0$.

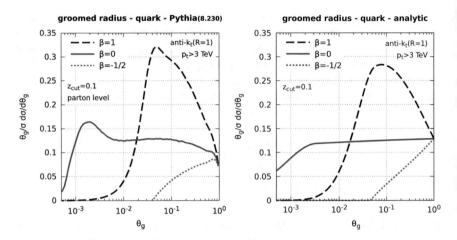

Fig. 9.2 The groomed radius distribution. The left plot is the result of a Pythia parton-level simulation and the right plot is the analytic results discussed in this chapter

more aggressive grooming, meaning a smaller jet area or a smaller sensitivity to pileup and the Underlying Event.

It is also worth noting that a few complications would arise if we wanted to extend Eq. (9.4) to full NLL accuracy. Since θ_g is only sensitive to the first emission being de-clustered that passes the SoftDrop condition, one could expect that it does not get any correction from multiple emissions at NLL. However, if one has multiple emissions at a similar angle and strongly-ordered in energy, which emission emission is de-clustered first will depend on the details of the Cambridge/Aachen clustering. This situation, which occurs only for $\beta > 0$, is reminiscent of the non-global and clustering logarithms discussed in Sects. 4.2.2 and 4.2.3. This type of effect has been discussed, for instance, in Refs. [204] and [65].

9.2   The z_g Distribution

We would like now to study the observable z_g. This immediately faces a difficulty: z_g is fixed by the first de-clustering of the jet that passes the SoftDrop condition. Because we are completely inclusive over the splitting angle θ_g we must take into account all possible values of θ_g including configurations where the two prongs become collinear. Indeed collinear splittings always pass the SoftDrop condition, if $\beta \geq 0$ (strictly speaking, soft-collinear emissions fail SoftDrop $\beta = 0$/mMDT). These configurations are not cancelled by the corresponding virtual corrections, for which z_g is undefined, and herald the fact that the observable is not IRC safe. At this point a possible approach would be to just stop this analysis because the observable we are dealing with does not respect the very basic set of properties set out in Chap. 2. However, we have just argued that z_g is a very interesting observable for jet substructure and therefore, we decide to be stubborn and push forward with our study. In order to do that, we must generalise the concept of IRC safety and introduce *Sudakov safety* [49].

Following [51], we introduce a general definition of Sudakov safety which exploits conditional probabilities. Let us consider an IRC unsafe observable u and a companion IRC safe observable s. The observable s is chosen such that its measured value regulates all singularities of u. That is, even though the probability of measuring u,

$$p(u) = \frac{1}{\sigma_0} \frac{d\sigma}{du}, \qquad (9.5)$$

is ill-defined at any fixed perturbative order, the probability of measuring u given s, $p(u|s)$, is finite at all perturbative orders, except possibly at isolated values of s e.g., $s = 0$ for definiteness. Given this companion observable s, we want to know whether $p(u)$ can be calculated in perturbation theory. Because s is IRC safe, $p(s)$ is well-defined at all perturbative orders and one can define the joint probability distribution

$$p(s, u) = p(s) \, p(u|s), \qquad (9.6)$$

which is also finite at all perturbative orders, except possibly at isolated values of s. To calculate $p(u)$, we can simply marginalise over s:

$$p(u) = \int ds \, p(s) \, p(u|s) . \qquad (9.7)$$

If $p(s)$ regulates all (isolated) singularities of $p(u|s)$, thus ensuring that the above integral is finite, then we deem u to be Sudakov safe.

Clearly, we cannot just evaluate $p(s)$ at fixed-order in the strong coupling, but we need some information about its all-order behaviour. If we consider the resummed distribution for the observable s, its distribution will exhibit a Sudakov form factor (hence the name Sudakov safety) that can make the integral in Eq. (9.7) convergent. In the case that one IRC safe observable is insufficient to regulate all singularities in u, we can measure a vector of IRC safe observables $\vec{s} = \{s_1, \ldots, s_n\}$. This gives a more general definition of Sudakov safety:

$$p(u) = \int d^n\vec{s} \, p(\mathbf{s}) \, p(u|\vec{s}) . \qquad (9.8)$$

Only if the vector of safe observables has a finite number of elements, then u is Sudakov safe. For example, particle multiplicity does not fall in this category as it would require an infinite number of safe observables to regulate the arbitrary number of soft/collinear splittings. Thus, particle multiplicity is neither IRC safe, nor Sudakov safe.

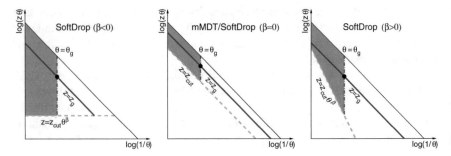

Fig. 9.3 Lund diagrams for the z_g distribution for three representative values of the SD angular exponent β. From left to right we have $\beta < 0$, $\beta = 0$ (mMDT) and $\beta > 0$. The dashed green line represents the edge of SD region The dot-dashed red line corresponds to emissions yielding a given groomed jet radius and the shaded red area is the vetoed area associated with the Sudakov suppression. Finally, the solid blue line corresponds to the requested value of z_g. Because we have to integrate over all possible values of θ_g, only the $\beta < 0$ case showed on the left exhibits IRC safety

We can now go back to the observable z_g and check whether it is Sudakov safe. To this purpose, we need to introduce a safe companion observable. The SoftDrop procedure itself suggests to use the groomed angle θ_g, which we have calculated in Eq. (9.3). Therefore, we imagine to measure a value of z_g, given a finite angular separation between the two prongs θ_g. This situation is illustrated by the Lund diagrams in Fig. 9.3. As usual, the dashed green line represents the edge of SoftDrop region. The black dot is the emission passing SoftDrop that provides z_g (solid blue line) and θ_g (dot-dashed red line). The shaded red area is the vetoed area associated with the Sudakov suppression for the groomed radius θ_g, i.e. it is the same as in Fig. 9.1. In order to obtain the z_g distribution, we have to integrate over all possible values of θ_g, which corresponds to all allowed positions for the dot-dashed red line. For $\beta < 0$, the area we swipe as we move the red dot-dashed line is bounded by the SoftDrop line in dashed green and it is therefore finite. In this case we expect IRC safety to hold. On the other hand, the $\beta = 0$ and $\beta > 0$ cases are remarkably different as the resulting area is unbounded. This situation is not IRC safe, but the Sudakov form factor for the groomed radius is enough to regulate (suppress) the resulting divergence. We have

$$\frac{1}{\sigma_0} \frac{d\sigma}{dz_g} = \int_0^1 d\theta_g \, p(\theta_g) \, p(z_g|\theta_g), \tag{9.9}$$

where $p(\theta_g)$ is the resummed distribution computed in the previous section, i.e. the derivative of Eq. (9.3), while the conditional probability is calculated at fixed-order in the strong coupling. In the collinear limit, it reads, for a jet of flavour $i = q, g$,

$$p(z_g|\theta_g) = \frac{P_{\text{sym},i}(z_g)\alpha_s(z_g\theta_g p_t R)}{\int_{z_{\text{cut}}\theta_g^\beta}^{1/2} dz \, P_{\text{sym},i}(z)\alpha_s(z \, \theta_g p_t R)} \Theta(z_g > z_{\text{cut}}\theta_g^\beta), \tag{9.10}$$

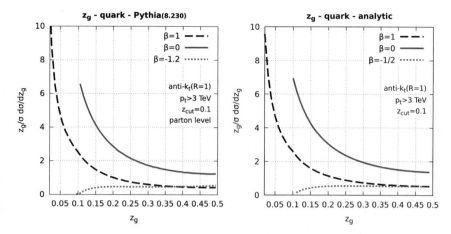

Fig. 9.4 The z_g distribution. The left plot is the result of a Pythia parton-level simulation and the right plot is the analytic results discussed in this chapter. We note that for $\beta < 0$ the observable is IRC safe, while for $\beta \geq 0$ it is only Sudakov safe

where $0 < z_g < 1/2$ and following the approach of Refs. [51, 192, 205], we have introduced a symmetrised notation of the splitting function

$$P_{\text{sym},i}(z) = P_i(z) + P_i(1 - z). \tag{9.11}$$

Crucially, the integral in Eq. (9.9) is finite for all values of β, provided we introduce a prescription for the Landau pole, and it can be evaluated numerically.[3] We also note that the z_g distribution in (9.9) is normalised to the ungroomed jet rate. This means that jets for which the SoftDrop procedure fails to find a two-prong structure, and so do not have a well-defined z_g, are still included in the normalisation of Eq. (9.9). This is obviously relevant for $\beta < 0$ where, even perturbatively, there is a finite probability for this to happen. For $\beta \geq 0$, this can also happen e.g. due to non-perturbative effects, or finite cut-offs in Monte Carlo simulations.

A comparison to parton-level Monte Carlo simulations is shown in Fig. 9.4, showing a remarkably good agreement given the collinear unsafety of the observable (for $\beta \geq 0$). What is perhaps more interesting is to try and understand explicitly the dominant behaviour of a Sudakov-safe observable. For this, we first work in the fixed-coupling limit. This means that, when evaluating Eq. (9.9), we can factor out $P_{\text{sym},i}(z_g)$ from Eq. (9.10) and z_g only enters as a phase-space constraint in the remaining integration. Next, we consider the soft limit. In this limit we can neglect hard-collinear splittings (i.e. the B terms) in the θ_g probability, and in Eq. (9.10) we can simplify the denominator by setting the upper bound of integration to 1 and set $P_{\text{sym},i}(z) \approx 2C_i/z$. The derivative of Eq. (9.3) needed for $p(\theta_g)$ brings a factor

---

[3]In practice, we have frozen the coupling at a scale $\mu_{\text{NP}} = 1$ GeV, cf. Appendix A.

$R'(\theta_g)$ which, with our assumptions, cancels the denominator of Eq. (9.10) up to a factor $C_i/(2\pi)$.[4] Writing $R(\theta_g)$ at fixed coupling, we are thus left with the following integration

$$\frac{1}{\sigma_0}\frac{d\sigma}{dz_g} = P_{\text{sym},i}(z_g)\frac{\alpha_s C_i}{\pi}\int_0^1 \frac{d\theta_g}{\theta_g}\exp\left[-\frac{\alpha_s C_i}{\pi\beta}\left(\log^2(z_{\text{cut}}\theta_g^\beta) - \log^2(z_{\text{cut}})\right)\right]$$
$$\Theta(z_{\text{cut}}\theta_g^\beta < z_g). \tag{9.12}$$

Let us first consider the case $\beta < 0$ for which $z_g > z_{\text{cut}}$ and we get[5]

$$\frac{1}{\sigma_0}\frac{d\sigma}{dz_g} \approx \sqrt{\frac{\alpha_s}{4|\beta|C_i}}e^{-\frac{\alpha_s C_i}{\pi|\beta|}\log^2(z_{\text{cut}})} \tag{9.13}$$

$$\left[\text{erfi}\left(\sqrt{\frac{\alpha_s C_i}{\pi|\beta|}}\log\left(\frac{1}{z_{\text{cut}}}\right)\right) - \text{erfi}\left(\sqrt{\frac{\alpha_s C_i}{\pi|\beta|}}\log\left(\frac{1}{z_g}\right)\right)\right]P_{\text{sym},i}(z_g),$$

where $\text{erfi}(x) = -i\,\text{erf}(ix)$ is the imaginary error function. For $\beta < 0$, z_g is an IRC-safe observable and, accordingly, the above result admits an expansion in powers of the strong coupling:

$$\beta < 0: \quad \frac{1}{\sigma_0}\frac{d\sigma}{dz_g} = \frac{\alpha_s}{\pi|\beta|}P_{\text{sym},i}(z_g)\log\left(\frac{z_g}{z_{\text{cut}}}\right)\Theta(z_g - z_{\text{cut}}) + \mathcal{O}(\alpha_s^2). \tag{9.14}$$

Moving now to $\beta > 0$, the evaluation of Eq. (9.12) gives

$$\frac{1}{\sigma_0}\frac{d\sigma}{dz_g} \approx \sqrt{\frac{\alpha_s}{4\beta C_i}}e^{\frac{\alpha_s C_i}{\pi\beta}\log^2(z_{\text{cut}})}\left[1-\text{erf}\left(\sqrt{\frac{\alpha_s C_i}{\pi\beta}}\log\left(\frac{1}{\min(z_g, z_{\text{cut}})}\right)\right)\right]P_{\text{sym},i}(z_g). \tag{9.15}$$

Although at first sight this looks similar to what was previously obtained, Eq. (9.15) (for positive β) shows a significantly different behaviour compared to Eq. (9.13) for negative β, as a direct consequence of the fact that z_g is only Sudakov safe for $\beta > 0$. Indeed, for $\beta > 0$, the distribution has the expansion

$$\beta > 0: \quad \frac{1}{\sigma_0}\frac{d\sigma}{dz_g} = \sqrt{\frac{\alpha_s}{4\beta C_i}}P_{\text{sym},i}(z_g) + \mathcal{O}(\alpha_s), \tag{9.16}$$

---

[4]This is easy to understand from a physical viewpoint: both $R'(\theta_g)$ and the denominator of Eq. (9.10) correspond to the probability for having a real emission passing the SoftDrop condition at a given θ_g.

[5]Note that the assumptions used in this book slightly differ from the ones originally used in [51].

and the presence of $\sqrt{\alpha_s}$ implies non-analytic dependence on α_s. This behaviour is associated with the "1" in the square bracket of Eq. (9.15), which can be traced back to the contribution from $\theta_g \to 0$ in Eq. (9.12), i.e. to the region where the observable is collinear unsafe (though Sudakov safe).

Finally, it is interesting to consider the specific case $\beta = 0$. At fixed coupling, $p(z_g|\theta_g)$ (Eq. (9.10)) is independent of θ_g and factors out of the θ_g integration in Eq. (9.9) to give

$$\beta = 0 : \quad \frac{1}{\sigma_0}\frac{d\sigma}{dz_g} = \frac{P_{\text{sym},i}(z_g)}{\int_{z_{\text{cut}}}^{1/2} dz\, P_{\text{sym},i}(z)}\Theta(z_g > z_{\text{cut}}). \tag{9.17}$$

It is not difficult to see that the $\beta = 0$ case does have a valid perturbative expansion in α_s, despite being α_s-independent at lowest order. This case is also only Sudakov safe, as the integration in Eq. (9.12) includes the collinear-unsafe region $\theta_g \to 0$. More generally, $\beta = 0$ marks the boundary between Sudakov-safe and IRC-safe situations. Eq. (9.17) has remarkable properties. Despite having being calculated in perturbative QCD, it exhibits a lowest-order behaviour which does not depend on the strong coupling, nor on any colour charge (in the small z_g limit). Instead, as anticipated, the distribution is essentially driven by the QCD splitting function, thus offering a unique probe of the dynamics of QCD evolution.

There exist now several examples of other Sudakov safe observables. These include ratios of angularities [49], the transverse momentum spectrum of a SoftDrop $\beta = 0$ (mMDT) jet [177], or equivalently the amount of energy which has been groomed away [50], and the pull angle [206], which is an observable that aims to measure colour-flow in a multi-jet event. Despite the very interesting results obtained thus far, the study of Sudakov safety is still in its infancy. Questions regarding the formal perturbative accuracy of the results, with related estimate of perturbative uncertainties, its dependence upon the choice of the safe companion, the inclusion of running coupling corrections, as well as the role of non-perturbative uncertainties are interesting theoretical aspects of perturbative QCD which are still actively investigated.

Chapter 10
Searches and Measurements with Jet Substructure

The previous chapters have focused on the theoretical and description of jet substructure variables, e.g. the jet mass, jet shapes and the classification of the jet-sourcing particles, together with some phenomenological studies performed with simulated data. In this chapter, we will give a brief overview of existing experimental performance studies, measurements and searches using jet substructure performed by ATLAS and CMS. As alluded to in Chaps. 3 and 4 all theoretical predictions of jet substructure observables can potentially deviate from experimental measurements for various reasons. For instance, theoretical calculations may fall short in capturing all relevant contributions or experimental effects, e.g. imperfect reconstruction of particle momenta, become important. Thus, it is of interest to see how well the theoretical predictions discussed in this book agree with experimental measurements and if found to be useful, in how far they can help in performing measurements of particle properties and searches for new physics. The large number of searches where jet substructure techniques are used in particular shows that it is a necessary ingredient in order to improve our understanding of nature. Here, we are not going to attempt to provide a comprehensive discussion of all searches and measurements performed by LHC experiments, but we will select and showcase results with close connection to the topics discussed before.

10.1 Tagging Performance Studies

Many taggers have been proposed have been proposed in the literature and we have reviewed a selection of them in Chap. 5. Often jet shapes or prong-finders are combined with other jet observables to perform a classification of the jet's initiating particle. Such a procedure can be augmented using machine-learning techniques to find the region of highest significance in the multi-dimensional parameter space of jet substructure observables. Different observables are used by ATLAS and CMS

© Springer Nature Switzerland AG 2019
S. Marzani et al., *Looking Inside Jets*, Lecture Notes in Physics 958,
https://doi.org/10.1007/978-3-030-15709-8_10

and their individual approaches have significantly evolved over the years. It is highly likely that the development of increasingly powerful classifiers, i.e. taggers, for jets will continue. Thus, in this brief review we will predominantly focus on ATLAS' and CMS' latest public performance comparisons.

ATLAS bases its W and top taggers on a set of techniques, rooted in jet shape observables, to determine a set of optimal cut-based taggers for use in physics analyses [207–209]. The first broad class of observables studied for classification rely on constituents of the trimmed jet to combine the topoclusters and tracks to a so-called combined jet mass m^{comb}. In addition to the jet mass, a set of jet shape observables are constructed: N-subjettiness ratios (τ_{21} and τ_{32}), splitting measures ($\sqrt{d_{12}}$ and $\sqrt{d_{23}}$), planar flow and energy correlation functions (C_i or D_i). Various subsets of these and similar observables are then combined in a boosted decision tree (BDT) or a deep neural network (DNN), see Table 10.1 for more details.

The performance of such multivariate BDT and DNN taggers is then compared to perturbative-QCD inspired taggers, i.e. the HEPTopTagger and the Shower Deconstruction tagger, using trimmed anti-k_t $R = 1.0$ fat jets. While the inputs to construct the observables of Table 10.1 consist of all jet constituents, the HEPTopTagger and Shower Deconstruction tagger are restricted to be used on calibrated Cambridge/Aachen subjets of finite size, i.e. $R_{\mathrm{subjet}} \geq 0.2$. Thus, ROC curves, as shown in Fig. 10.1, have to be taken with a grain of salt, as systematic uncertainties of the input objects have not been propagated consistently into the performance curves.[1] However, in particular for highly boosted top quarks, see Fig. 10.1, the combination of multiple jet shape observables shows a very strong tagging performance over the entire signal efficiency range.

CMS [220, 221] takes a similar approach to W boson and top quark tagging as ATLAS. CMS uses a subset of the observables of Table 10.1, and extends it by including Qjet volatility [152] and b-tagging[2] in their performance analysis. In addition to the Shower Deconstruction tagger, an updated version of the HEPTopTagger (V2) and the CMS top tagger are included in the comparison. The results of Fig. 10.2 (left) show that the performance of individual observables and taggers can vary a lot, with Shower Deconstruction performing best in the signal efficiency region of $\varepsilon_S \leq 0.7$. However, when various tagging methods are combined in a multivariate approach, Fig. 10.2 (right), their performance become very similar and the potential for further improvements seems to saturate for the scenario at hand. For W tagging, see Fig. 10.3, CMS combines several jet shape observables using either a naive Bayes classifier or a Multilayer Perceptron (MLP) neural network discriminant.

---

[1] Systematic [217, 218] and theoretical [219] uncertainties can be taken into account in the performance evaluation of a neural net classifier by adding an adversarial neural network.

[2] For jets, b-tagging is meant to separate jets originating from a b quark from light-quark and gluon jets. b-tagging algorithms are using the fact that B hadrons decay with a displaced vertex together with a list of variables included in a BDT or neural network (with details depending on the experiment).

Table 10.1 A summary of the set of observables that were tested for W-boson and top-quark tagging for the final set of DNN and BDT input observables [207]

W boson tagging

| | DNN test groups | | | | | | | | | Inputs | |
	1	2	3	4	5	6	7	8	9	BDT	DNN
m^{comb}	o	o	o	o	o	o	o	o	o	o	o
p_t	o	o	o	o	o	o	o	o	o	o	o
e_3	o	o				o			o		
C_2			o	o	o		o	o	o	o	o
D_2			o	o	o		o	o	o	o	o
τ_1	o	o	o			o			o	o	
τ_2	o	o	o			o				o	
τ_3											
τ_{21}			o	o	o	o	o	o	o	o	o
τ_{32}			o	o	o	o	o	o	o	o	o
R_2^{FW}					o	o	o	o	o	o	o
$\mathcal{P}$			o	o	o	o	o	o	o	o	o
a_3			o	o	o	o	o	o	o	o	o
A			o	o	o	o	o	o	o	o	o
z_{cut}			o	o	o	o	o	o	o	o	o
$\sqrt{d_{12}}$		o				o	o	o	o	o	
$\sqrt{d_{23}}$							o	o	o	o	
$KtDR$		o				o	o	o	o	o	
Q_w						o	o	o	o	o	o

Top quark tagging

| | DNN test groups | | | | | | | | | Inputs | |
	1	2	3	4	5	6	7	8	9	BDT	DNN
m^{comb}	o	o	o	o	o	o	o	o	o	o	o
p_t	o	o	o	o	o	o	o	o	o	o	o
e_3			o				o	o	o	o	o
C_2					o	o					o
D_2	o	o	o		o	o					o
τ_1					o	o			o		o
τ_2	o	o		o	o				o	o	o
τ_3				o			o				o
τ_{21}	o	o	o		o	o	o	o	o	o	o
τ_{32}	o	o	o		o	o	o	o	o	o	o
R_2^{FW}											
$\mathcal{P}$											
a_3											
A											
z_{cut}											
$\sqrt{d_{12}}$					o		o	o	o	o	o
$\sqrt{d_{23}}$					o		o	o	o	o	o
$KtDR$											
Q_w					o	o	o		o	o	o

p_t and m^{comb} are the transverse momentum of the jet and the combined jet mass [210]. e_3, C_2 and D_2 are energy correlation ratios [143, 211]. τ_l and τ_{ij} are N-subjettiness variables and ratios respectively. R_2^{FW} is a Fox-Wolfram moment [212]. Splitting measures are denoted z_{cut}, $\sqrt{d_{12}}$ and $\sqrt{d_{23}}$ [149, 213]. The planar flow variable $\mathcal{P}$ is defined in [214] and the angularity a_3 in [215]. Definitions can be found for aplanarity A [216], $KtDR$ [57] and Q_w [149]

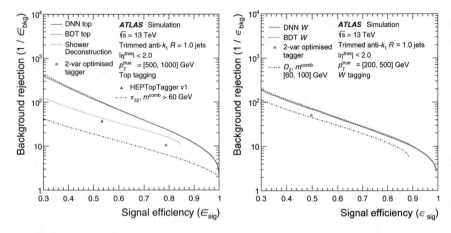

Fig. 10.1 Top quark (left) and W boson (right) tagging efficiencies for various tagging approaches used by ATLAS [207]

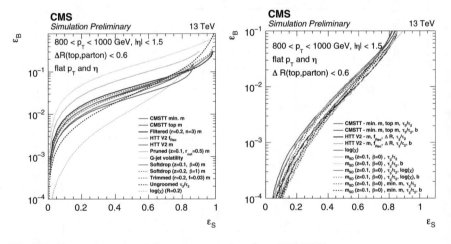

Fig. 10.2 Top quark tagging performance comparison from CMS [220]

When comparing to individual jet shape observables, such as N-subjettiness ratios or Qjet volatility, mild improvements can be achieved.

The discrimination between quark and gluon-initiated jets can have profound phenomenological implications. A large class of processes associated with the production of new particles have a strong preference to result in quarks, e.g. the production and subsequent decay of squarks in the Minimal Supersymmetric Standard Model, while Standard Model QCD backgrounds are more likely to result in gluon-initiated jets. Thus, the ability to separate these two classes of jets reliably could boost our sensitivity in finding new physics. However, as discussed at length in Chap. 7, the discrimination between a jet that was initiated by a gluon from a jet that

Fig. 10.3 W boson tagging performance comparison from CMS [221]

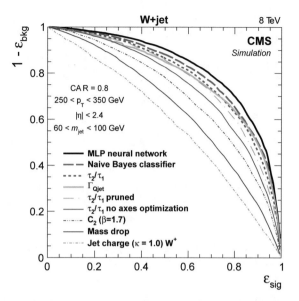

was initiated by a quark is subtle. Consequently, sophisticated observables which attempt to exploit small features between quarks and gluon jets can potentially be sensitive to limited experimental resolution and experimental uncertainties in the construction of the jet constituents. In their performance studies, ATLAS [222] and CMS [223] aim to exploit the differences in the radiation profiles between quarks and gluons using observables such as the number of charged tracks n_{trk}, calorimeter w_{cal} or track width w_{trk} with

$$w = \frac{\sum_i p_{T,i} \times \Delta R(\mathrm{i}, \mathrm{jet})}{\sum_i p_{T,i}} , \qquad (10.1)$$

where i runs either over the calorimeter energy clusters to form w_{cal} or over the charged tracks for w_{trk}. Further observables are the track-based energy-energy-correlation (EEC) angularities

$$\mathrm{ang}_{\mathrm{EEC}} = \frac{\sum_i \sum_j p_{T,i} \, p_{T,j} \, (\Delta R(i, j))^\beta}{(\sum_i p_{T,i})^2} , \qquad (10.2)$$

where the index i and j run over the tracks associated with the jet, with $j > i$, and β is a tunable parameter, the jet minor angular opening σ_2 of the p_t^2-weighted constituents distribution in the lego plane and the jet fragmentation distribution $p_T D$, defined as

$$p_T D = \frac{\sqrt{\sum_i p_{T,i}^2}}{\sum_i p_{T,i}}, \qquad (10.3)$$

where i runs over all jet constituents.

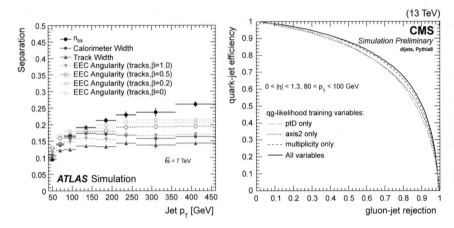

Fig. 10.4 ATLAS (left) and CMS (right) studies of quark/gluon discrimination. The plots are taken from, respectively, Refs. [222] and [223]

ATLAS results for quark-gluon tagging [222, 224] are reported in Fig. 10.4 on the left, in terms of the variable "separation", which is defined as:

$$\text{Separation} = \frac{1}{2} \int \frac{(p_q(x) - p_g(x))^2}{p_q(x) + p_g(x)} dx \qquad (10.4)$$

where $p_q(x)$ and $p_g(x)$ are normalised distributions of the variables used for discrimination between quark and gluon jets. Both experiments achieve a good separation between quark and gluon jets for the observables used and the p_t-windows studied. For example, CMS achieves for a 50% quark jet acceptance a rejection of roughly 90% of gluon jets.

10.2   Measurements of Jet Observables

Various jet observables discussed in Chaps. 5–9 have been measured by ATLAS and CMS. In the following we will discuss a selection of the measurements performed for these observables and we will focus on measurements for large-R jets.

10.2.1   Jet Mass

The mass of a jet is one of the most basic observables associated with a jet. As such, it was discussed in great detail in Chaps. 4 and 6, with and without the application of various grooming methods to the jet. As the mass is sensitive to the energy distribution in the jet, it can also be thought of as a jet-shape observable.

ATLAS [225, 226] and CMS [227, 228] have both measured the mass of jets under various conditions. In [225] ATLAS has measured the jet mass, amongst other jet shape and jet substructure observables, in pp collisions at a centre-of-mass energy of 7 TeV.

The SoftDrop mass has been measured in [226] and [228] by ATLAS and CMS, respectively. After requiring a jet with $p_t > 600$ GeV and imposing the dijet topology cut $p_{T,1}/p_{T,2} < 1.5$, ATLAS runs the soft-drop algorithm on the two leading jets in the events. Three different values of $\beta \in 0, 1, 2$ are considered, while the value on the z_{cut} is fixed at 0.1. Then the dimensionless ratio $m^{\text{soft drop}}/p_t^{\text{ungroomed}}$ is constructed and shown in Fig. 10.5 in the lower right panel. The measured data is in good agreement with various theoretical predictions, including resummed analytic calculations and full event generators. CMS selects similar event kinematics for this measurement, but fixes $\beta = 0$. In Fig. 10.5 in the upper right and upper left panel the groomed and ungroomed jet mass, measured by CMS at 13 TeV centre-of-mass energy, is shown respectively. Data is

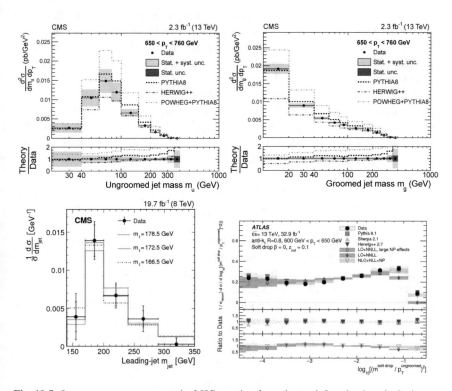

Fig. 10.5 Jet mass measurements at the LHC, starting from the top left and going clockwise, we have: plain jet mass by CMS [228], SoftDrop (mMDT) jet mass by CMS [228], SoftDrop mass measurement by ATLAS [226], and top jet mass by CMS [227]

compared with theory predictions from Pythia8, Herwig++ and Powheg+Pythia8, showing significant differences between the three event generators. While the overall normalisation of the cross sections predicted by the event generators is quite different, with Pythia8 being closest to data, the shape of the theoretically predicted distributions agree well with data. Thus, when the distributions are normalised to the total cross section, the difference between data and all three theory predictions is small.

The precision with which a boosted top quark's mass can be measured by analysing a fat jet is a crucial parameter for many tagging algorithms. In [227] CMS purifies the final state with respect to semi-leptonic $t\bar{t}$ events and reconstructs Cambridge/Aachen $R = 1.2$ fat jets with $p_t > 400$ GeV. The mass of the leading fat jet is sown in the lower left panel of Fig. 10.5. No special grooming procedure has been used, yet the measured jet mass agrees well with the physical top mass.

10.2.2   Jet Charge

The energy deposits and tracks associated with a jet can originate from dozens of charged particles, depending on the size and on transverse momentum of the jet. If charged particles become too soft, e.g. $p_t \ll 1$ GeV, they can curl up in the magnetic field of the detector and might not even be measurable in the calorimetry or the tracker. Thus, it is useful to define the jet charge as a p_t-weighted sum of the charge of the jet constituents. As the number of charged particles amongst the jet constituents is neither an infrared-safe nor a perturbatively calculable quantity, experimental measurements of these observables have to be compared to fitted hadronisation models included in full event generators. A natural and common definition for the jet charge is [229, 230]

$$Q_J = \frac{1}{(p_{T,J})^\kappa} \sum_{i \in \text{tracks}} q_i \, (p_{T,i})^\kappa \,, \tag{10.5}$$

where i runs over all tracks associated with jet J. q_i is the measured charge of track i with associated transverse momentum $p_{T,i}$, and κ is a free regularisation parameter.[3] In this definition the charge associated with individual tracks, i.e. individual charged particles, is weighted by their transverse momentum. That way Q_J is less sensitive to experimental and theoretical uncertainties. ATLAS and CMS both find good agreement between theoretical predictions and data over a large range of transverse momenta of the jets, when calculating their average charge, see Fig. 10.6.

---

[3]There are alternative definitions of jet charge. For examples and how their theoretical prediction compares to experimental measurements, see [230].

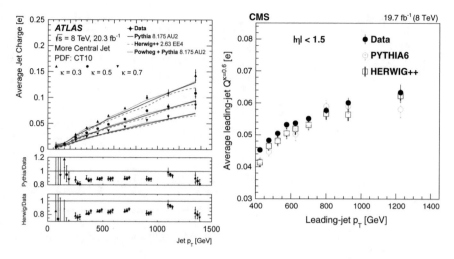

Fig. 10.6 Measurements of the jet charge by ATLAS [229] (left) and CMS [230] (right)

10.2.3 Splitting Functions

The momentum sharing z_g between the two subjets that pass the SoftDrop condition was introduced in Sect. 9.2. The variable z_g can be taken as a proxy of the "most important" partonic splitting in the jet evolution and thus its distribution is governed by the QCD splitting functions. A measurement[4] of the z_g distribution in pp collisions, using CMS open data, was reported in [192, 205]. Using data obtained during LHC's heavy-ion runs, CMS has studied z_g in PbPb and pp collisions [231]. A measurement in of z_g in PbPb collisions reflects how the two colour-charged partons produced in the first splitting propagate through the quark-gluon plasma, thereby probing the role of colour coherence of the jet in the medium. In the pp case all particle-flow anti-k_t jets with $R = 0.4$ and $p_{t,j} > 80$ GeV were recorded. To identify the hard prongs of a jet and to remove soft wide-angle radiation, SoftDrop grooming is applied to the jets with $\beta = 0$ and $z_{cut} = 0.1$. Figure 10.7 shows the comparison of z_g between the measured CMS data and the theoretical predictions from Pythia6, Pythia8 and Herwig++, including a full simulation of detector effects. While in general good agreement is observed, both Pythia simulations have a slightly steeper z_g distribution than the data, whereas Herwig++ shows an opposite trend.

---

[4]Note that we refer here to observables that have not been unfolded. Thus, a comparison of data to theoretical predictions requires the knowledge of detector effects on the reconstructed observable.

Fig. 10.7 The groomed
momentum sharing z_g
measured by CMS during
LHC's heavy-ion runs in pp
collisions [231]

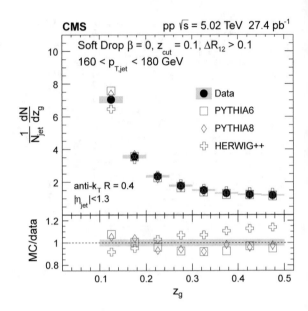

10.3   Search for Boosted Higgs Boson in the SM

The possibility to search for the Standard Model Higgs boson in the decay to $b\bar{b}$ at
the LHC using jet substructure techniques gave the field of jet physics a tremendous
boost [5]. The projected sensitivity for a discovery of the Higgs boson with only
~ 30 fb^{-1}, however, requires a centre-of-mass energy of 14 TeV for LHC proton-
proton collisions. Due to technical issues of the LHC to reach its design energy of
14 TeV during Runs I and II, the decay of a Higgs boson into a $b\bar{b}$-pair was never a
contender to contribute to its discovery. Still, the measurement of the Higgs boson
coupling to bottom quarks, while notoriously difficult, is of crucial importance as it
is a dominant contributor to the total width of the Higgs boson, which in turn affects
the branching ratios of all available decay modes.

   CMS performed an inclusive search for a Higgs boson decaying to $b\bar{b}$ pair, which
is expected to result in an anti-k_t $R = 0.8$ jet, with $p_t \geq 450$ GeV [232]. The main
experimental challenge originates in the large cross section for background multijet
events at low jet mass. To increase the sensitivity for the reconstruction of the Z and
Higgs boson SoftDrop grooming is applied to the jet before two- and three-point
generalised energy correlation functions are exploited to determine how consistent
a jet is with having a two-prong structure. While a peak is clearly visible for the
reconstruction of the Z boson, Fig. 10.8 (top left) shows that the sensitivity to the
Higgs boson still remains weak. However, it is already possible to set a limit on large
signal-strength modifications to the production of either resonance, see Fig. 10.8
(top right).

   ATLAS has provided a similar measurement with an increased data set of $\mathcal{L} =
80.5$ fb^{-1} [233]. To select the event, an anti-k_t $R = 1$ fat jet with $p_t \geq 480$ GeV is

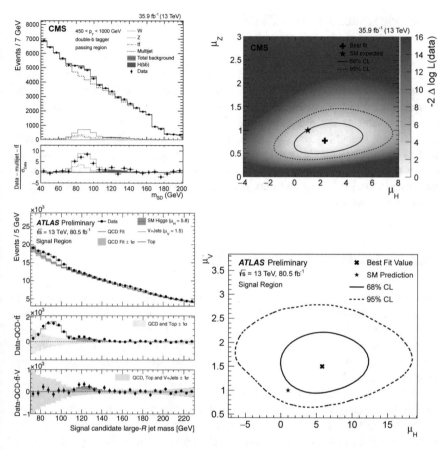

Fig. 10.8 Searches for boosted Higgs boson decaying into $b\bar{b}$. The plots show the invariant mass distribution and the signal-strength modification, top for CMS [232], bottom for ATLAS [233]

required. ATLAS is not showing the soft-drop groomed mass of the fat jet, but the invariant mass of trimmed jets. After subtracting the rather large QCD background a clear excess around the Higgs mass of $m_H = 125$ GeV is observed, see Fig. 10.8 (lower left panel). Small excesses around the Higgs and Z masses are indicative of an enhanced signal strength compared to the Standard Model predicted cross sections. Thus, ATLAS central value for the fit, allowing the signal strength for the V+jets and H+jets independently to float, is above the Standard Model value for either process, see Fig. 10.8 (lower right panel). Yet, ATLAS and CMS 95% exclusion contours both still contain the Standard Model value.

10.4   Searches for New Physics

The kinematic situation outlined at the beginning of this book, cf Fig. 1.1, is
common to many scenarios where the Standard Model is extended by heavy degrees
of freedom. If such degrees of freedom descend from a model that addresses the
hierarchy problem of the Higgs boson, they are likely to couple to the top quark
and the bosons of the electroweak sector of the Standard Model, which in turn
have a large branching ratio into jets. The conversion of energy from the heavy
particle's rest mass into kinetic energy of the much lighter electroweak resonances
causes them to be boosted in the lab frame. Thus, searches for new physics using jet
substructure methods applied to fat jets can be amongst the most sensitive ways to
probe new physics.

10.4.1   Resonance Decays into Top Quarks

Top-tagging is the most active playground for the development of jet substructure
classification techniques. A top jet has a rich substructure, providing several handles
to discriminate it from the large QCD backgrounds, and due to the top quark's
short lifetime its dynamics are to a large degree governed by perturbative physics.
Thus, ATLAS and CMS have performed searches using a large variety of top-
reconstruction techniques.

   New physics scenarios that are the focus of ATLAS and CMS searches contain
models with extra dimensions or extended gauge groups, which give rise to heavy
Z' bosons, Kaluza-Klein gluons g_{KK} and spin-2 Kaluza-Klein gravitons G_{KK}. The
hadronic activity, and hence the tagging efficiency, depends on the quantum numbers
of the heavy decaying resonance, in particular its colour charge [234]. These three
resonances provide interesting benchmark points which can arise in many classes of
new physics models.

   While ATLAS [235] separates between a resolved and a boosted analysis in the
semi-leptonic top-decay channel, i.e. with one top decaying leptonically ($t \rightarrow bvl^{+}$)
and the other one hadronically ($t \rightarrow bjj$), CMS [236] focuses on the boosted
regime but also considers the dileptonic and purely hadronic top decay modes. For
the purpose of these notes, we are mostly interested in the boosted semi-leptonic
top-decay mode, which suffers less from large dijet backgrounds, yet providing a
larger signal cross section than the dileptonic channel. ATLAS varies the resonance
masses for the colour-singlet and colour-octet bosons with spin 1 or spin 2 between
0.4 to 5 TeV and respectively their width between 1% and 30%. To reconstruct
the hadronic top, a large-R jet is formed using the anti-k_t algorithm with radius
parameter $R = 1.0$. This jet is trimmed to mitigate the effects of pileup and
underlying event, using $R_{\mathrm{sub}} = 0.2$ and $f_{\mathrm{cut}} = 0.05$. The resulting jets are required
to have $p_t > 300$ GeV and $|\eta| < 2.0$. Such jets are then identified as top-tagged

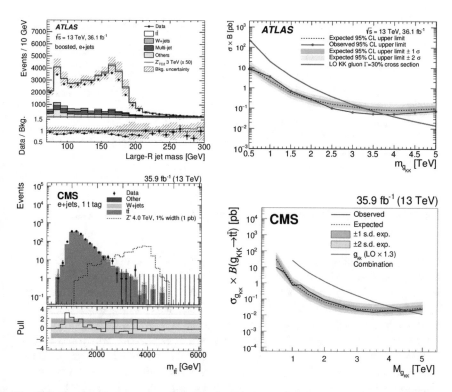

Fig. 10.9 Searches for heavy resonances that involve top tagging from ATLAS [235], on the top, and CMS [236] on the bottom

using the N-subjettiness ratio τ_{32} and an algorithm based on the invariant mass of the jet. The signal efficiency for this algorithm is found to be 80%.

For the same task CMS uses somewhat smaller anti-k_t $R = 0.8$ jets. These jets receive pileup per particle identification (PUPPI) [174] corrections. The top tagging algorithm then only considers jets with $p_t > 400$ GeV to ensure a collimated decay of the top quark. The top tagging algorithm then includes a grooming step, performed with SoftDrop $\beta = 0$, i.e. mMDT, algorithm, with $z_{\text{cut}} = 0.1$ and $R_0 = 0.8$ and a cut on the N-subjettiness ratio τ_{32}. The SoftDrop mass is then required to be close enough to the true top mass, i.e. $105 < m_{\text{SD}} < 210$ GeV and τ_{32} must be less than 0.65.

The reconstruction techniques applied show a very good agreement between the measured data and the Monte-Carlo predicted pseudo-data, Fig. 10.9 (left panels). With only 36 fb^{-1}, depending on the resonance's couplings and width, heavy resonances decaying into top quarks can be excluded up to mass of 3.5 TeV, Fig. 10.9 (right panels). For such large masses jet-substructure methods are not optional. Without using the internal structure of jets the QCD-induced dijet backgrounds would overwhelm the signal.

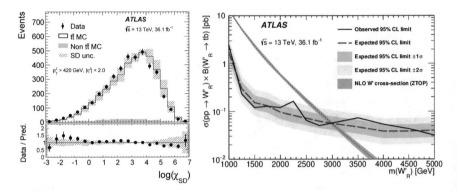

Fig. 10.10 Search for a W′ performed by ATLAS [237] using shower deconstruction

In models where the Z′ arises from to a SU(N) gauge group, it will be accompanied by a W′. ATLAS [237] has performed searches for heavy W′ decaying into a hadronic top and a bottom quark, i.e. $W' \rightarrow t\bar{b} \rightarrow q\bar{q}b\bar{b}$. This search is somewhat more intricate than the searches for decays into two top quarks as there are fewer handles to suppress the backgrounds. Thus, ATLAS uses the shower deconstruction top tagging algorithm, discussed in Sect. 5.6.1, which has a strong rejection power of QCD jets while maintaining a large signal efficiency. ATLAS finds a working point of the tagger with 50% signal efficiency and a background rejection factor of 80, thus improving the signal-to-background ratio by a factor 40, for anti-k_t $R = 1.0$ jets with $p_t > 450$ GeV. Applied to $t\bar{t}$ events the shower deconstruction algorithm shows very good agreement between data and Monte-Carlo simulated pseudo-data, Fig. 10.10 (left). After applying the top tagger and reconstructing the final state in the search for a W′, ATLAS can exclude masses up to 3 TeV, Fig. 10.10 (right). The sensitivity does not depend on whether the W′ couples to left or right-handed quarks.

10.4.2 Resonance Decays into Higgs and Gauge Bosons

Currently, most of the resonance searches into electroweak bosons that are using jet substructure techniques are focusing either on heavy resonances decaying into Higgs bosons, with subsequent decay into bottom quarks [238–240], or decays into W and Z bosons, with subsequent decay into quarks [241, 242]. The new physics scenarios studied range from the decay of a Kaluza-Klein excitation of the graviton in the bulk Randall-Sundrum model with a warped extra dimension [243], over the decay of a CP-even heavy Higgs boson, as present in two-Higgs double models (2HDM) [244], to a heavy scalar from a triplet-Higgs model, e.g. the so-called Georgi-Machacek model [245]. In general one assumes a heavy resonance in the range $m_X \gtrsim 500$ GeV is produced, with a short lifetime. While the spin of the resonance could be in principle studied by reconstructing and analysing the decay planes of the quark pairs [246, 247], at this point such attempts are not being made and the separation between signal and background, after reconstructing the electroweak bosons, relies entirely on the presence of a bump in their invariant mass distribution. Thus, the width in combination with the mass of the decaying resonance is of great importance for its discovery or exclusion.

To reconstruct the Higgs boson pairs from a heavy resonance decay ATLAS [238, 248] considers a resolved and a boosted analysis. In the boosted case, Higgs bosons are selected by requiring that two large-R, i.e. anti-k_t $R = 1.0$ with $p_t \geq 250$ GeV, have each two b-tags, and that the leading fat jet has in addition $p_t \geq 350$ GeV. The backgrounds are derived from Monte-Carlo simulations. For a resonance width of $\Gamma = 1$ GeV, in combination with the resolved analysis, this results in an 95% C.L. exclusion for a bulk Randall-Sundrum graviton with coupling value $k/\tilde{M}_{PL} = 2$, where $\tilde{M}_{PL}$ is the reduced Planck mass, of $500 \leq m_{G^*_{KK}} \leq 990$ GeV, see Fig. 10.11 (left). With the integrated luminosity used in this analysis, the resolved analysis is more sensitive than the boosted analysis up to $m_{G^*_{KK}} \leq 1100$ GeV.

CMS [240] in a search at $\sqrt{s} = 13$ TeV with 35.9 fb^{-1} aims for the exclusion of heavier masses and instead uses anti-k_t $R = 0.8$ jets with $p_t \geq 350$ GeV. The resulting fat jet is groomed using the SoftDrop algorithm with $z = 0.1$ and $\beta = 0$ (mMDT). The groomed jet mass is required to have $105 \leq m_{sd} \leq 135$ GeV. To suppress QCD backgrounds further, the N-subjettiness algorithm is used, requiring $\tau_{21} < 0.55$. Eventually, each of the jets is double-b tagged, which has the largest impact on the backgrounds. After searching for a bump in the invariant mass spectrum of the two fat jets, CMS obtains a 95% C.L. exclusion for a bulk radion with mass $970 < m_R < 1400$ GeV, see Fig. 10.11 (right).

The decay of a heavy resonance into gauge bosons is a frequent feature of many extensions of the Standard Model. For example, the aforementioned bulk graviton could decay into a pair of W or Z bosons, or a heavy gauge boson of an additional or extended gauge group, a so-called W' or Z', can decay into the pairs of Standard Model gauge bosons. In [241, 242] ATLAS and CMS have both observed a small

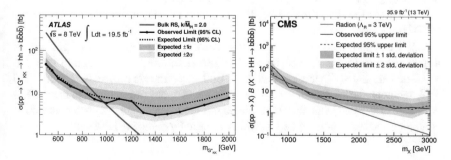

Fig. 10.11 Example of exclusions limits from searches of resonances decaying into two Higgs bosons, the ATLAS search of Ref. [238] is shown on the left, while the CMS search of Ref. [240] on the right

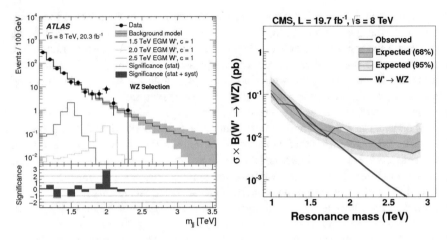

Fig. 10.12 Invariant mass distribution of the reconstructed gauge bosons as measured by ATLAS [241], on the left, and the exclusion limit for a W' decaying into WZ set by CMS [242] on the right

excess in dijet final states, where each jet was W/Z tagged. The excess resided at a similar invariant mass range of $m_{jj} \sim 2$ TeV, but was slightly more significant in the ATLAS analysis. In this search, at $\sqrt{s} = 8$ TeV, ATLAS selected two fat jets with Cambridge/Aachen algorithm $R = 1.2$, with a minimal transverse momentum of $p_t \geq 540$ GeV. The reconstruction of the gauge bosons relied on a combination of a jet-mass cut around the masses of the weak gauge bosons and grooming techniques, where a modified version of the BDRS reconstruction technique [5] was employed,[5] a method initially designed for the reconstruction of a Higgs boson with $p_{t,H} \geq 200$ GeV. Using this approach, the mass resolution of the reconstructed gauge boson is not good enough to discriminate between W and Z bosons. Eventually, to improve on the separation of signal and background, cuts were applied on the momentum ratios of subjets, the number of charged particles within a subjet and the mass of the reconstructed gauge bosons. After recombining the four-momenta of the two reconstructed gauge bosons an excess was observed in the mass range $1.9 \leq m_{VV} \leq$ 2.1 TeV over the data-driven (fitted) background estimate, mainly driven by the QCD background, see Fig. 10.12.

CMS reconstructed the W and Z bosons by applying the pruning algorithm as a groomer and tagger for the fat jet. To further improve the separation between W/Z bosons and QCD jets τ_{21} was used. Both experiments find an excess at ~ 1.9 TeV, whereas the excess in ATLAS with 2.8σ is more pronounced than in CMS with 1.8σ. Both experiments have updated this search using different reconstruction strategies and with more statistics, which eventually dampened the excess strongly [250, 251].

---

[5] As discussed in [249], the way the BDRS approach was modified in the search by ATLAS could result in shaping the m_{VV} distribution in the region of 2 TeV, where the excess was observed.

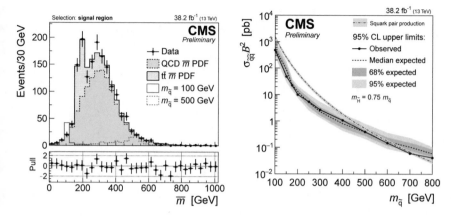

Fig. 10.13 Search for a squark decaying into four quarks (via a Higgsino) performed by CMS [253]. The plot on the right show the average invariant mass of the fat jets and the exclusion limits

10.4.3   Resonance Decays into New Particles

If a heavy resonance decays via electroweak resonances of the Standard Model into quarks or gluons the masses of the Standard Model particles themselves provide an important handle to separate signal from background. This task is however complicated if the intermediate resonances are not known, e.g. when a squark decays into four quarks through an intermediate Higgsino with a hadronic R-parity-violating coupling [252]. CMS has performed searches for squarks via such a decay mode [253] and ATLAS a similar study in [254].

Without knowing the mass of the Higgsino, CMS in [253] uses the fact that a squark decays into a four-pronged object to discriminate signal from background. In order to capture as many of the final-state constituents of the squark decay products as possible, large Cambridge/Aachen ($R = 1.2$) jets are formed. These jets are analysed using the N-subjettiness ratios, requiring $\tau_{43} < 0.8$ and $\tau_{42} < 0.5$ for each jet. As pair production of the squarks is assumed, the masses of the reconstructed fat jets should not be too asymmetric, i.e. $|m_1 - m_2|/(m_1 + m_2) < 0.1$. After defining the average jet mass as $\bar{m} = (m_1 + m_2)/2$ CMS finds very good agreement between the theoretically predicted background cross sections and the measured data, see Fig. 10.13 (left). This allows to set an exclusion limit in this channel requiring squark masses to be $m_{\tilde{q}} > 720$ GeV, when assuming that the Higgsino mass is $m_{\tilde{H}} = 0.75m_{\tilde{q}}$, Fig. 10.13 (right).

Chapter 11
Take-Home Messages and Perspectives

Since many facets and applications of jet substructure have been covered in this book, it is useless to try and summarise them all individually. Instead, in this concluding chapter, we will briefly summarise the main lessons we have learned from about a decade of jet substructure studies and from the aspects covered in this book.

The first observation is that jet substructure has been a great success, both from a theoretical viewpoint and from an experimental viewpoint. It took only a few years for the initial idea of looking at the internal dynamics of jets to grow and develop a myriad of new tools, opening doors to explore all sorts of new physics domains. Furthermore, as searches and measurements probe larger and larger energy scales, boosted-object and jet substructure algorithms are increasingly relied upon. In particular, at a possible future circular hadron collider with $\sqrt{s}$ as large as 100 TeV, boosted jets would be almost omnipresent.

In practice, jet substructure tools are rooted in the theory of strong interactions. The first generation of substructure techniques were designed based on core concepts and features of QCD: a QCD jet is usually made of a single hard core accompanied with soft particles corresponding to soft-gluon radiation, while boosted massive objects decay into several hard prongs accompanied by further soft radiation (at smaller angles if the initial particle is colourless). Key techniques, many of which still in use today, have been developed starting from these fundamental observations, allowing to establish jet substructure as a powerful and promising field. A few years later, the introduction of a new generation of substructure tools was made possible by a better understanding of the QCD dynamics inside jets using analytic techniques. This first-principle approach has allowed for a more fine-grained description of the underlying physics, which seeded either simpler and cleaner tools (e.g. the modified MassDropTagger and SoftDrop) or tools with improved performance (e.g. the D_2 energy-correlation functions or dichroic ratios), all under good theoretical control.

© Springer Nature Switzerland AG 2019
S. Marzani et al., *Looking Inside Jets*, Lecture Notes in Physics 958,
https://doi.org/10.1007/978-3-030-15709-8_11

One of the key features repeatedly appearing when studying jet substructure from first principles in QCD is the necessity of a trade-off between performance and robustness. Here, by performance, we mean the discriminating power of a tool when extracting a given signal from the QCD background, and by robustness we mean the ability to describe the tool from perturbative QCD, i.e. being as little sensitive as possible to model-dependent effects such as hadronisation, the Underlying Event, pileup or detector effects, all of which likely translate into systematic uncertainties in an experimental analysis. This trade-off has been seen on multiple occasions throughout this book. When designing new substructure techniques, we therefore think that it is helpful to keep in mind both these aspects.

In this context, it was realised that some tools like SoftDrop or the modified MassDropTagger are amenable to precise calculations in perturbative QCD, while maintaining small hadronisation and Underlying Event corrections. This is particularly interesting since jet substructure tools are often sensitive to a wide range of scales—between the TeV scale down to non-perturbative scales—offering an almost unique laboratory for QCD studies. It has opened new avenues for future jet substructure studies. A typical example is a potential for an extraction of the strong coupling constant from substructure measurements (see e.g. Ref. [122]), but other options include the improvement of Monte Carlo parton showers, measurements of the top mass, or simply a better control over QCD background for new physics searches.

Because of its potential for interesting Standard Model measurements across a wide range of scales, jet substructure has also recently found applications in heavy-ion collisions. One of the approaches to study the quark-gluon plasma is by analysing how high-energy objects are affected by their propagation through it. The LHC is the first collider where jets are routinely used for this type of studies and an increasing interest for jet substructure observables has been seen very recently in the heavy-ion community. This will for sure be an important avenue in the future of jet substructure, including the development of specific observables to constraint the properties of the quark-gluon plasma and their study in QCD.

The analysis of cosmic ray interactions is a further area of research where jet substructure techniques were introduced to study the detailed structure of complicated objects [255, 256]. Ultra-high-energy cosmic rays, e.g. protons, can produce interactions with very high momentum transfer between when they scatter of atoms of Earth's atmosphere. Such interactions produce a collimated high-multiplicity shower of electrons, photons and muons. Their spacial distribution and penetration depth can be analysed to inform the nature of the incident particle and interaction in the collision. It is likely that in the near future, with the increased interest in so-called beam-dump experiments, more ideas are going to be introduced where jet substructure techniques can become of importance.

Finally, one should also expect the future to deliver its fair share of new tools for searches and measurements. We believe that there are two emblematic directions worth exploring. An obvious direction is the one of machine-learning tools. This is an increasingly hot topic in the jet substructure community and one should expect it to continue growing in importance. In the context of the first-principle understanding used throughout this book, one should highlight that it is important to keep in mind that applying machine-learning techniques to jet substructure is not just a problem for computer scientists. These algorithms are to a large extent dealing with QCD and therefore a good control of the QCD aspects of jet substructure is crucial. Several examples of this have appeared very recently—like QCD-aware networks [257], energy-flow polynomials and networks [146, 147] or the Lund jet plane [65]—and we should definitely expect more in the future. One can even imagine to extend concepts developed for jets to be applied to the full event, i.e. a full-information approach to study the whole radiation profile of an event. This could maximise the sensitivity of collider experiments in searches for new physics.

The second direction we want to advocate for is the development of additional tools which are theory-friendly, i.e. that are under analytical control and are amenable for precision calculations. As shown in this book, basic substructure tools have now been understood from first-principles, including the main physics aspects responsible for the trade-off between performance and robustness. However, modern boosted jet taggers involve several of these tools in order to maximise performance (cf. our discussion in Chap. 5).

We think that new tools offering a combination of grooming, prong-finding and radiation constraints will always be of great value. Compared to a deep-learning-based tool, this might show a small loss in performance, but it would offer the advantage of a better control of its behaviour across a wide range of processes and studies. One of the key ingredients here is that these new tools should remain as simple as possible to facilitate their calibration in an experimental context, hopefully resulting in small systematic uncertainties. This would make them usable for the precision programme at the LHC, including both measurements and searches. From an analytic perspective, achieving precision for such substructure algorithms will also require further developments in resummation techniques and fixed-order (amplitude) calculations, where many promising results have already been obtained recently.

All this being said, we hope that we have conveyed the idea that jet substructure has been a fascinating field for almost a decade, with an ever-growing range of applications. Over this time-span, the field has managed to stay open to new ideas and new approaches. One should therefore expect more exciting progress in the years to come. We therefore hope that this book gives a decent picture of the state of the field in early 2019 and will constitute a good introduction for newcomers to the field.

If you ain't boostin' you ain't living
¡Boostamos!
(https://www.youtube.com/watch?v=zsUIDOn6nJk)

Appendix A
Details of Analytic Calculations

In this appendix we detail the analytic calculations that we have to perform in order to obtain the resummed exponents discussed in the main text. As an example we consider the plain jet mass distribution discussed in Chap. 4. The generalisation to other jet substructure observables merely adds additional phase-space constraints, yielding longer expressions without changing the steps of the calculation. It is left as an exercise for the interested reader.

We therefore consider the resummed expression Eq. (4.16) and we focus on the resummed exponent (focusing here on a quark-initiated jet, although similar results can trivially be obtained for gluon-initiated jets)

$$R(\rho) = \int_\rho^1 \frac{d\rho'}{\rho'} \int_{\rho'}^1 dz \, P_q(z) \frac{\alpha_s(\sqrt{z\rho'}R\mu)}{2\pi}, \tag{A.1}$$

where μ is the hard scale of the process, i.e. $\mu = \frac{Q}{2}$ for electron-positron collisions or $\mu = p_t$ for proton-proton collision, while as usual R is the jet radius, and $\rho = \frac{m^2}{\mu^2 R^2}$. For the above expression to capture the resummed exponent to NLL accuracy in the small-R limit, we need to make sure that

- the running of the coupling is considered at two loops, i.e. with β_0 and β_1:

$$\alpha_s(k_t) = \frac{\alpha_s(R\mu)}{1 + \tilde{\lambda}} \left[1 - \alpha_s(\mu R) \frac{\beta_1}{\beta_0} \frac{\log(1 + \tilde{\lambda})}{1 + \tilde{\lambda}} \right], \quad \tilde{\lambda} = 2\alpha_s(R\mu)\beta_0 \log\left(\frac{k_t}{R\mu}\right), \tag{A.2}$$

where the β function coefficients β_0 and β_1 are

$$\beta_0 = \frac{11C_A - 2n_f}{12\pi}, \quad \beta_1 = \frac{17C_A^2 - 5C_A n_f - 3C_F n_f}{24\pi^2}. \tag{A.3}$$

© Springer Nature Switzerland AG 2019
S. Marzani et al., *Looking Inside Jets*, Lecture Notes in Physics 958,
https://doi.org/10.1007/978-3-030-15709-8

- the splitting function is considered at one loop;
- the coupling is considered in the CMW scheme (or equivalently the soft contribution to the two-loop splitting function is included), cf. Eq. (4.18).

As a warm up, let us first evaluate the above integral to LL where, we can limit ourselves to the soft limit of the splitting function and to the one-loop approximation for the running coupling. We have (with $\lambda' = \alpha_s \beta_0 \log(\rho')$ and $\lambda'' = \alpha_s \beta_0 \log(z)$)

$$
\begin{aligned}
R^{(\text{LL})} &= \int_\rho^1 \frac{d\rho'}{\rho'} \int_{\rho'}^1 \frac{dz}{z} \frac{\alpha_s(\sqrt{z\rho'}R\mu)C_F}{\pi} \\
&= \frac{\alpha_s C_F}{\pi} \int_\rho^1 \frac{d\rho'}{\rho'} \int_{\rho'}^1 \frac{dz}{z} \frac{1}{1 + \alpha_s \beta_0 \log(z\rho')} \\
&= \frac{C_F}{\alpha_s \pi \beta_0^2} \int_{-\frac{\lambda}{2}}^0 d\lambda' \int_{\lambda'}^0 d\lambda'' \frac{1}{1 + \lambda' + \lambda''} \\
&= \frac{C_F}{2\pi \beta_0^2 \alpha_s} \left[(1-\lambda)\log(1-\lambda) - 2\left(1 - \frac{\lambda}{2}\right)\log\left(1 - \frac{\lambda}{2}\right)\right], \\
&= \frac{C_F}{2\pi \beta_0^2 \alpha_s} \left[W(1-\lambda) - 2W\left(1 - \frac{\lambda}{2}\right)\right],
\end{aligned}
\tag{A.4}
$$

where $\alpha_s \equiv \alpha_s(R\mu)$ is the $\overline{\text{MS}}$ coupling, $\lambda = 2\alpha_s \beta_0 \log\left(\frac{1}{\rho}\right)$, and $W(x) = x\log(x)$. The above result can be then easily recast in the form of the f_1 function Eq. (4.20), which appears in the expression for the resummed exponent Eq. (4.19).

Next, we consider the inclusion of the hard-collinear contribution. For this we have to include regular part of the splitting function. Thus, we have to evaluate the following integral:

$$
\begin{aligned}
\delta R^{(\text{hard-collinear})} &= \int_\rho^1 \frac{d\rho'}{\rho'} \int_{\rho'}^1 \frac{dz}{z} \left[P_q(z) - \frac{2}{z} \right] \frac{\alpha_s(\sqrt{z\rho'}R\mu)}{\pi} \\
&= \frac{2C_F \alpha_s}{\pi} \int_\rho^1 \frac{d\rho'}{\rho'} \int_{\rho'}^1 dz \left[-1 + \frac{z}{2} \right] \frac{1}{1 + \alpha_s \beta_0 \log(z\rho')}.
\end{aligned}
\tag{A.5}
$$

When evaluating the expression above to NLL we can make the further simplifications that, since we are working in the hard-collinear limit, we can set $z = 1$ in the running-coupling contribution. We are left with an integral over z with no logarithmic enhancement so, up to power corrections in ρ, we can safely set the lower limit of integration to $z = 0$. The two integrals decouple and we find

$$
\delta R^{\text{(hard-collinear)}} = \frac{C_F \alpha_s}{\pi} \int_0^1 dz \left[-1 + \frac{z}{2} \right] \int_\rho^1 \frac{d\rho'}{\rho'} \frac{1}{1 + \alpha_s \beta_0 \log(\rho')}
$$

$$
= -\frac{C_F}{\pi \beta_0} B_q \log\left(1 - \frac{\lambda}{2} \right), \tag{A.6}
$$

with

$$
B_q = \int_0^1 dz \left[\frac{P_q(z)}{2 C_F} - \frac{1}{z} \right] = \int_0^1 dz \left[-1 + \frac{z}{2} \right] = -\frac{3}{4}, \tag{A.7}
$$

already defined in Eq. (4.10). Note that for a gluon-initiated jet one should instead use the gluon splitting function, Eq. (2.28), which includes a contribution from $g \to gg$ splitting and one from $g \to q\bar{q}$ splitting:

$$
B_g = \int_0^1 dz \left[\frac{P_g(z)}{2 C_A} - \frac{1}{z} \right] = -\frac{11 C_A - 2 n_f}{12 C_A}. \tag{A.8}
$$

Since hard-collinear splittings often have a large numerical impact and are relatively easy to include, one often works in the *modified LL approximation* where one includes the LL contribution $R^{(\text{LL})}$ as well as hard-collinear splittings, $\delta R^{\text{(hard-collinear)}}$.

Before moving on to the other NLL contributions to the Sudakov exponent, we would like to comment on an alternative way to achieve modified leading logarithmic accuracy and include the "B-term" in the LL expressions. We note that if we replace the actual splitting function by any other expressions which behaves like $\frac{2 C_i}{z}$ at small z and reproduces the correct B_i term in Eqs. (A.7) and (A.8), we would then recover the same modified-LL behaviour. In particular, we can use

$$
P_i^{\text{(modified-LL)}}(z) = \frac{2 C_i}{z} \Theta\left(z < e^{B_i} \right). \tag{A.9}
$$

This is equivalent to imposing a cut on z in the LL integrals. For example, Eq. (A.4) would become

$$
R^{\text{(modified-LL)}} = \int_\rho^1 \frac{d\rho'}{\rho'} \int_{\rho'}^{e^{B_i}} \frac{dz}{z} \frac{\alpha_s (\sqrt{z \rho'} R \mu) C_F}{\pi} \tag{A.10}
$$

$$
= \frac{C_i}{2 \pi \beta_0^2 \alpha_s} \left[W(1 - \lambda) - 2W \left(1 - \frac{\lambda + \lambda_B}{2} \right) + W(1 - \lambda_B) \right],
$$

with $\lambda_B = -2 \alpha_s \beta_0 B_i$. It is straightforward to show that if we expand this to the first non-trivial order in λ_B, one indeed recovers $R^{(\text{LL})} + \delta R^{\text{(hard-collinear)}}$. This approach is what we have adopted for most of the results and plots presented in this book.

Coming back to the full NLL accuracy for the resummed exponent, we also have to consider the contribution of the two-loop running coupling:

$$
\begin{aligned}
\delta R^{(2\text{-loop})} &= -\frac{\alpha_s^2 C_i}{\pi} \frac{\beta_1}{\beta_0} \int_\rho^1 \frac{d\rho'}{\rho'} \int_{\rho'}^1 \frac{dz}{z} \frac{\log(1 + \alpha_s \beta_0 \log(z\rho'))}{(1 + \alpha_s \beta_0 \log(z\rho'))^2} \\
&= -\frac{C_i \beta_1}{\pi \beta_0^3} \int_{\frac{-\lambda}{2}}^0 d\lambda' \int_{\lambda'}^0 d\lambda'' \frac{\log(1 + \lambda' + \lambda'')}{(1 + \lambda' + \lambda'')^2} \\
&= \frac{C_i \beta_1}{2\pi \beta_0^3} \left[\log(1 - \lambda) - 2\log\left(1 - \frac{\lambda}{2}\right) \right. \\
&\qquad\qquad \left. + \frac{1}{2}\log^2(1 - \lambda) - \log^2\left(1 - \frac{\lambda}{2}\right) \right],
\end{aligned}
\tag{A.11}
$$

which provides the β_1 contribution to the NLL function f_2 defined in Eq. (4.21).

Finally, to NLL accuracy we also have to include the two-loop contribution to the splitting function in the soft limit. Because this contribution is universal it can be also expressed as a redefinition of the strong coupling, which give rise to the so-called CMW scheme Eq. (4.18). Thus, we have to evaluate the following integral

$$
\begin{aligned}
\delta R^{(\text{CMW})} &= \frac{2 C_i K}{4\pi^2} \int_\rho^1 \frac{d\rho'}{\rho'} \int_{\rho'}^1 \frac{dz}{z} \alpha_s^2(\sqrt{z\rho'} R\mu) \\
&= \frac{C_i K}{2\pi^2 \beta_0^2} \int_{\frac{-\lambda}{2}}^0 d\lambda' \int_{\lambda'}^0 d\lambda'' \frac{1}{(1 + \lambda' + \lambda'')^2} \\
&= \frac{C_i K}{4\pi^2 \beta_0^2} \left[2\log\left(1 - \frac{\lambda}{2}\right) - \log(1 - \lambda) \right],
\end{aligned}
\tag{A.12}
$$

where the coupling in the first line can be evaluated at the one-loop accuracy since higher-order corrections would be beyond NLL. This contribution is the K term in the NLL function f_2 defined in Eq. (4.21).

The expressions in this appendix allow us to capture the global part of resummed exponent to NLL, in the small-R limit. Had we decided to include finite R correction, we would have considered also soft emissions at finite angles, not just from the hard parton in the jet but from all dipoles of the hard scattering process (see for instance Sect. 4.3 and Ref. [93]). Furthermore, we remind the reader that, as discussed in Chap. 4, in order to achieve full NLL accuracy, one needs to consider non-global logarithms as well as potential logarithmic contributions originating from the clustering algorithm which is used to define the jet.

Finally, we note that the above expressions exhibit a singular behaviour at $\lambda = 1$ and $\lambda = 2$. These singularities originate from the Landau pole of the perturbative QCD coupling and they signal the breakdown of perturbation theory. In phenomenological applications of analytic calculations this infrared region is dealt by introducing a particular prescription. For instance, one could imagine to freeze the coupling below a non-perturbative scale $\mu_{NP} \simeq 1$ GeV

$$\bar{\alpha}_s(\mu) = \alpha_s(\mu)\Theta(\mu - \mu_{NP}) + \alpha_s(\mu_{NP})\Theta(\mu_{NP} - \mu). \tag{A.13}$$

Other prescriptions are also possible. For example, in Monte Carlo simulations, the parton showers is typically switched off at a cutoff scale and the hadronisation model then fills the remaining phase-space.

With the prescription Eq. (A.13), the above expressions for the Sudakov exponent are modified at large λ. For completeness, we give the full expressions resulting from the more tedious but still straightforward integrations. To this purpose, it is helpful to introduce $W(x) = x\log(x)$, $V(x) = \frac{1}{2}\log^2(x) + \log(x)$, and $\lambda_{fr} = 2\alpha_s\beta_0\log\left(\frac{\mu R}{\mu_{NP}}\right)$. For $\lambda < \lambda_{fr}$, i.e. $\rho > \frac{\mu_{NP}}{R\mu}$, we find

$$R^{(NLL)}(\lambda) = R^{(\text{modified-LL})} + \delta R^{(2\text{-loop})} + \delta R^{(CMW)} \tag{A.14}$$

$$= \frac{C_i}{2\pi\alpha_s\beta_0^2}\left[W(1-\lambda) - 2W\left(1 - \frac{\lambda+\lambda_B}{2}\right) + W(1-\lambda_B)\right]$$

$$+ \frac{C_i\beta_1}{2\pi\beta_0^3}\left[V(1-\lambda) - 2V\left(1 - \frac{\lambda+\lambda_B}{2}\right) + V(1-\lambda_B)\right]$$

$$- \frac{C_i K}{4\pi^2\beta_0^2}\left[\log(1-\lambda) - 2\log\left(1 - \frac{\lambda+\lambda_B}{2}\right) + \log(1-\lambda_B)\right],$$

in agreement with Eqs. (A.10)–(A.12) above. We note that the above expressions have included the B term using the trick of Eq. (A.9) for all terms including the two-loop and CMW corrections. In these terms, one can set $\lambda_B = 0$ at NLL accuracy. Although keeping these contribution has the drawback of introducing uncontrolled subleading corrections, it comes with the benefit of providing a uniform treatment of hard-collinear splitting which places the endpoint of all the terms in the resummed distribution at $\lambda = \lambda_B$.

For $\lambda_{fr} < \lambda < 2\lambda_{fr}$, i.e. $\left(\frac{\mu_{NP}}{R\mu}\right)^2 < \rho < \frac{\mu_{NP}}{R\mu}$, we start being sensitive to the freezing of the coupling at μ_{NP}. In this case, we find

$$R^{(NLL)}(\lambda) = \frac{C_i}{2\pi\alpha_s\beta_0^2} \tag{A.15}$$

$$\left[(1-\lambda)\log(1-\lambda_{fr}) - 2W\left(1 - \frac{\lambda+\lambda_B}{2}\right) + W(1-\lambda_B) + \frac{1}{2}\frac{(\lambda-\lambda_{fr})^2}{1-\lambda_{fr}}\right]$$

$$+ \frac{C_i\beta_1}{2\pi\beta_0^3}\left[\frac{1}{2}\log^2(1-\lambda_{fr}) + \frac{1-\lambda}{1-\lambda_{fr}}\log(1-\lambda_{fr}) - 2V\left(1 - \frac{\lambda+\lambda_B}{2}\right)\right.$$

$$\left. + V(1-\lambda_B) - \frac{\lambda-\lambda_{fr}}{1-\lambda_{fr}} - \frac{1}{2}\frac{(\lambda-\lambda_{fr})^2}{(1-\lambda_{fr})^2}\log(1-\lambda_{fr})\right]$$

$$-\frac{C_i K}{4\pi^2 \beta_0^2}\left[\log(1-\lambda_{\mathrm{fr}}) - 2\log\left(1 - \frac{\lambda+\lambda_B}{2}\right)\right.$$

$$\left. + \log(1-\lambda_B) - \frac{\lambda-\lambda_{\mathrm{fr}}}{1-\lambda_{\mathrm{fr}}} - \frac{1}{2}\frac{(\lambda-\lambda_{\mathrm{fr}})^2}{(1-\lambda_{\mathrm{fr}})^2}\right].$$

Finally, for $\lambda > 2\lambda_{\mathrm{fr}}$, i.e. $\rho < \left(\frac{\mu_{\mathrm{NP}}}{R\mu}\right)^2$, we have

$$R^{(\mathrm{NLL})}(\lambda) = R^{(\mathrm{NLL})}(\lambda_{\mathrm{fr}})$$

$$+ \frac{C_i}{2\pi\alpha_s\beta_0^2}\frac{(\lambda-\lambda_B)^2 - 2(\lambda_{\mathrm{fr}}-\lambda_B)^2}{4(1-\lambda_{\mathrm{fr}})} \qquad (A.16)$$

$$\left[1 - \frac{\alpha_s\beta_1}{\beta_0}\frac{\log(1-\lambda_{\mathrm{fr}})}{1-\lambda_{\mathrm{fr}}} + \frac{\alpha_s K}{2\pi}\frac{1}{1-\lambda_{\mathrm{fr}}}\right].$$

For all the analytic plots in this paper, we have used $\alpha_s(M_Z) = 0.1265$ (following the value used for the (one-loop) running coupling in Pythia8 with the Monash 2013 tune), freezing α_s at $\mu_{\mathrm{NP}} = 1$ GeV and used five active massless flavours. Note finally that (modified)-LL results only include one-loop running coupling effects.

Appendix B
Details of Monte Carlo Simulations

In this appendix, we provide the details of the parton-shower Monte Carlo simulations presented throughout this book.

For all the results shown in Chaps. 6, 7 and 9, we have used the Pythia 8 generator [178, 258] (version 8.230) with the Monash 2013 tune [179]. The analytic results are always compared to Monte Carlo results at parton level, where both hadronisation and the Underlying Event have been switched off. The "hadron level" corresponds to switching on hadronisation but keeping multi-parton interactions off, while the "hadron+UE" level includes both hadronisation and the Underlying Event. In the last two cases, B-hadrons have been kept stable for simplicity.[1] For the samples labelled as "quark jets", we have used Pythia's dijet hard processes, keeping only the $qq \to qq$ matrix elements. Similarly the "gluon jet" samples keep only the $gg \to gg$.

For all studies, jet reconstruction and manipulations are performed using Fast-Jet [56, 77] (version 3.3). Our studies include all the jets above the specified p_t cut and with $|y| < 4$. Substructure tools which are not natively included in FastJet are available from `fastjet-contrib`.

In the case of the discrimination between boosted W jets and QCD jets in Chap. 8, we have used the same samples as those used in the initial Les-Houches 2017 Physics at TeV colliders workshop. These use essentially the same generator settings as above, but now only up to the two hardest jets with $|y| < 2.5$ are kept.

---

[1]Except for the groomed jet mass study in Sect. 6.3 where B-hadron decays are enabled.

© Springer Nature Switzerland AG 2019
S. Marzani et al., *Looking Inside Jets*, Lecture Notes in Physics 958,
https://doi.org/10.1007/978-3-030-15709-8

References

1. M.H. Seymour, Tagging a heavy Higgs boson, in *ECFA Large Hadron Collider Workshop: Proceedings.2*, Aachen, 4–9 October 1990 (1991), pp. 557–569
2. M.H. Seymour, Searches for new particles using cone and cluster jet algorithms: a comparative study. Z. Phys. **C62**, 127–138 (1994)
3. M.H. Seymour, The average number of subjets in a hadron collider jet. Nucl. Phys. **B421**, 545–564 (1994)
4. J. Butterworth, B. Cox, J.R. Forshaw, WW scattering at the CERN LHC. Phys. Rev. **D65**, 096014 (2002). [hep-ph/0201098]
5. J.M. Butterworth, A.R. Davison, M. Rubin, G.P. Salam, Jet substructure as a new Higgs search channel at the LHC. Phys. Rev. Lett. **100**, 242001 (2008). [0802.2470]
6. A. Abdesselam et al., Boosted objects: a probe of beyond the standard model physics. EPHJA, C71, 1661.2011 **C71**, 1661 (2011). [1012.5412]
7. A. Altheimer, S. Arora, L. Asquith, G. Brooijmans, J. Butterworth et al., Jet substructure at the Tevatron and LHC: new results, new tools, new benchmarks. J. Phys. **G39**, 063001 (2012). [1201.0008]
8. A. Altheimer, A. Arce, L. Asquith, J. Backus Mayes, E. Bergeaas Kuutmann et al., Boosted objects and jet substructure at the LHC. Report of BOOST2012, held at IFIC Valencia, 23rd–27th of July 2012. Eur. Phys. J. **C74**, 2792 (2014). [1311.2708]
9. D. Adams et al., Towards an understanding of the correlations in jet substructure. Eur. Phys. J. **C75**, 409 (2015). [1504.00679]
10. A.J. Larkoski, I. Moult, B. Nachman, Jet substructure at the large Hadron collider: a review of recent advances in theory and machine learning (2017). 1709.04464
11. L. Asquith et al., Jet substructure at the large Hadron collider : experimental review (2018). 1803.06991
12. D.E. Soper, M. Spannowsky, Combining subjet algorithms to enhance ZH detection at the LHC. J. High Energy Phys. **08**, 029 (2010). [1005.0417]
13. L.G. Almeida, R. Alon, M. Spannowsky, Structure of fat jets at the tevatron and beyond. Eur. Phys. J. **C72**, 2113 (2012). [1110.3684]
14. STAR collaboration, K. Kauder, Measurement of the shared momentum fraction z_g using jet reconstruction in p+p and Au+Au collisions with STAR. Nucl. Phys. **A967**, 516–519 (2017). [1704.03046]
15. S. Catani, B.R. Webber, G. Marchesini, QCD coherent branching and semiinclusive processes at large x. Nucl. Phys. **B349**, 635–654 (1991)
16. S. Catani, L. Trentadue, G. Turnock, B.R. Webber, Resummation of large logarithms in e^+e^- event shape distributions. Nucl. Phys. **B407**, 3–42 (1993)

© Springer Nature Switzerland AG 2019
S. Marzani et al., *Looking Inside Jets*, Lecture Notes in Physics 958,
https://doi.org/10.1007/978-3-030-15709-8

17. G. Luisoni, S. Marzani, QCD resummation for hadronic final states. J. Phys. **G42**, 103101 (2015) . [1505.04084]
18. T. Becher, A. Broggio, A. Ferroglia, *Introduction to Soft-Collinear Effective Theory*. Lecture Notes in Physics, vol. 896 (2015), pp. 1–206. [1410.1892]
19. M.E. Peskin, D.V. Schroeder, *An Introduction to Quantum Field Theory* (Westview, Boulder, 1995)
20. M.D. Schwartz, *Quantum Field Theory and the Standard Model* (Cambridge University Press, Cambridge, 2014)
21. J. Collins, *Foundations of Perturbative QCD*. Cambridge Monographs on Particle Physics, Nuclear Physics, and Cosmology (Cambridge University Press, New York, 2011)
22. R.K. Ellis, W.J. Stirling, B.R. Webber, *QCD and Collider Physics*, vol. 8 (Cambridge University Press, Cambridge, 1996)
23. J. Campbell, J. Huston, F. Krauss, *The Black Book of Quantum Chromodynamics: A Primer for the LHC Era* (Oxford University Press, Oxford, 2018)
24. J.R. Forshaw, A. Kyrieleis, M. Seymour, Super-leading logarithms in non-global observables in QCD. J. High Energy Phys. **0608**, 059 (2006) . [hep-ph/0604094]
25. J. Forshaw, A. Kyrieleis, M. Seymour, Super-leading logarithms in non-global observables in QCD: colour basis independent calculation. J. High Energy Phys. **0809**, 128 (2008). [0808.1269]
26. S. Catani, D. de Florian, G. Rodrigo, Space-like (versus time-like) collinear limits in QCD: is factorization violated? J. High Energy Phys. **1207**, 026 (2012). [1112.4405]
27. J.R. Forshaw, M.H. Seymour, A. Siodmok, On the breaking of collinear factorization in QCD. J. High Energy Phys. **1211**, 066 (2012). [1206.6363]
28. G. Soyez, Pileup mitigation at the LHC: a theorist's view (2018). 1801.09721
29. C. Anastasiou, C. Duhr, F. Dulat, F. Herzog, B. Mistlberger, Higgs boson gluon-fusion production in QCD at three loops. Phys. Rev. Lett. **114**, 212001 (2015) . [1503.06056]
30. F.A. Dreyer, A. Karlberg, Vector-boson fusion Higgs production at three loops in QCD. Phys. Rev. Lett. **117**, 072001 (2016) . [1606.00840]
31. F. Bloch, A. Nordsieck, Note on the radiation field of the electron. Phys. Rev. **52**, 54 (1937)
32. T. Kinoshita, Mass singularities of feynman amplitudes. J. Math. Phys. **3**, 650 (1962)
33. T. Lee, M. Nauenberg, Degenerate systems and mass singularities. Phys. Rev. **B133**, 1549 (1964)
34. S. Catani, M. Ciafaloni, G. Marchesini, Non-cancelling infrared divergences in QCD coherent state. Nucl. Phys. **B264**, 588–620 (1986)
35. S. Catani, M.H. Seymour, A general algorithm for calculating jet cross sections in NLO QCD. Nucl. Phys. **B485**, 291–419 (1997). [hep-ph/9605323]
36. S. Catani, M.H. Seymour, The dipole formalism for the calculation of QCD jet cross sections at next-to-leading order. Phys. Lett. **B378**, 287–301 (1996) . [hep-ph/9602277]
37. Y. Dokshitzer, G. Marchesini, Soft gluons at large angles in hadron collisions. J. High Energy Phys. **0601**, 007 (2006) . [hep-ph/0509078]
38. Y. Dokshitzer, G. Marchesini, Hadron collisions and the fifth form-factor. Phys. Lett. **B631**, 118–125 (2005) . [hep-ph/0508130]
39. B.I. Ermolaev, V.S. Fadin, Log-log asymptotic form of exclusive cross-sections in quantum chromodynamics. JETP Lett. **33**, 269–272 (1981)
40. A.H. Mueller, On the multiplicity of hadrons in QCD jets. Phys. Lett. **B104**, 161–164 (1981)
41. A. Bassetto, M. Ciafaloni, G. Marchesini, A.H. Mueller, Jet multiplicity and soft gluon factorization. Nucl. Phys. **B207**, 189 (1982)
42. G.F. Sterman, S. Weinberg, Jets from quantum chromodynamics. Phys.Rev.Lett. **39**, 1436 (1977)
43. G.F. Sterman, Mass divergences in annihilation processes. 1. Origin and nature of divergences in cut vacuum polarization diagrams. Phys. Rev. **D17**, 2773 (1978)
44. G.F. Sterman, Mass divergences in annihilation processes. 2. Cancellation of divergences in cut vacuum polarization diagrams. Phys. Rev. **D17**, 2789 (1978)

45. G.F. Sterman, Zero mass limit for a class of jet related cross-sections. Phys. Rev. **D19**, 3135 (1979)
46. J.C. Collins, D.E. Soper, The theorems of perturbative QCD. Ann. Rev. Nucl. Part. Sci. **37**, 383–409 (1987)
47. H.-M. Chang, M. Procura, J. Thaler, W.J. Waalewijn, Calculating track-based observables for the LHC. Phys. Rev. Lett. **111**, 102002 (2013) . [1303.6637]
48. H.-M. Chang, M. Procura, J. Thaler, W.J. Waalewijn, Calculating track thrust with track functions. Phys. Rev. **D88**, 034030 (2013) . [1306.6630]
49. A.J. Larkoski, J. Thaler, Unsafe but calculable: ratios of angularities in perturbative QCD. J. High Energy Phys. **1309**, 137 (2013) . [1307.1699]
50. A.J. Larkoski, S. Marzani, G. Soyez, J. Thaler, Soft drop. J. High Energy Phys. **1405**, 146 (2014). [1402.2657]
51. A.J. Larkoski, S. Marzani, J. Thaler, Sudakov safety in perturbative QCD. Phys. Rev. **D91**, 111501 (2015) . [1502.01719]
52. A.J. Larkoski, D. Neill, J. Thaler, Jet shapes with the broadening axis. J. High Energy Phys. **1404**, 017 (2014). [1401.2158]
53. P.E.L. Rakow, B.R. Webber, Transverse momentum moments of Hadron distributions in QCD jets. Nucl. Phys. **B191**, 63–74 (1981)
54. G.P. Salam, Towards jetography. Eur. Phys. J. **C67**, 637–686 (2010). [0906.1833]
55. J.E. Huth et al., Toward a standardization of jet definitions, in *1990 DPF Summer Study on High-energy Physics: Research Directions for the Decade (Snowmass 90) Snowmass, Colorado, June 25–July 13, 1990* (1990), pp. 0134–136
56. M. Cacciari, G.P. Salam, G. Soyez, FastJet user manual. Eur. Phys. J. **C72**, 1896 (2012). [1111.6097]
57. S. Catani, Y.L. Dokshitzer, M.H. Seymour, B.R. Webber, Longitudinally-invariant $k_\perp$-clustering algorithms for hadron–hadron collisions. Nucl. Phys. **B406**, 187–224 (1993)
58. S.D. Ellis, D.E. Soper, Successive combination jet algorithm for hadron collisions. Phys. Rev. **D48**, 3160–3166 (1993) . [hep-ph/9305266]
59. JADE collaboration, W. Bartel et al., Experimental studies on multijet production in $e^+ e^-$ annihilation at PETRA energies. Z. Phys. **C33**, 23 (1986)
60. JADE collaboration, S. Bethke et al., Experimental investigation of the energy dependence of the strong coupling strength. Phys. Lett. **B213**, 235–241 (1988)
61. Y.L. Dokshitzer, G. Leder, S. Moretti, B. Webber, Better jet clustering algorithms. J. High Energy Phys. **9708**, 001 (1997) . [hep-ph/9707323]
62. M. Wobisch, T. Wengler, Hadronization corrections to jet cross-sections in deep inelastic scattering (1998). hep-ph/9907280
63. M. Cacciari, G.P. Salam, G. Soyez, The anti-k(t) jet clustering algorithm. J. High Energy Phys. **04**, 063 (2008). [0802.1189]
64. M. Cacciari, G.P. Salam, G. Soyez, The catchment area of jets. J. High Energy Phys. **04**, 005 (2008) . [0802.1188]
65. F.A. Dreyer, G.P. Salam, G. Soyez, The Lund jet plane. J. High Energy Phys. **12**, 064 (2018). [1807.04758]
66. CDF collaboration, F. Abe et al., The topology of three jet events in $\bar{p}p$ collisions at $\sqrt{s} = 1.8$ TeV. Phys. Rev. **D45**, 1448–1458 (1992)
67. G.C. Blazey et al., Run II jet physics, in *QCD and Weak Boson Physics in Run II. Proceedings*, Batavia, March 4–6, June 3–4, November 4–6, 1999 (2000), pp. 47–77. hep-ex/0005012
68. D0 collaboration, V.M. Abazov et al., Measurement of the inclusive jet cross section in $p\bar{p}$ collisions at $\sqrt{s} = 1.96$ TeV. Phys. Rev. **D85**, 052006 (2012). [1110.3771]
69. G.P. Salam, G. Soyez, A practical seedless infrared-safe cone jet algorithm. J. High Energy Phys. **05**, 086 (2007) . [0704.0292]
70. CMS collaboration, Commissioning of the particle-flow event reconstruction with the first LHC collisions recorded in the CMS detector, CMS-PAS-PFT-10-001
71. CMS collaboration, Commissioning of the particle-flow reconstruction in minimum-bias and jet events from pp collisions at 7 TeV (2010). CMS-PAS-PFT-10-002

72. Particle Data Group collaboration, M. Tanabashi et al., Review of particle physics. Phys. Rev. **D98**, 030001 (2018)
73. ATLAS collaboration, G. Aad et al., Single hadron response measurement and calorimeter jet energy scale uncertainty with the ATLAS detector at the LHC. Eur. Phys. J. **C73**, 2305 (2013). [1203.1302]
74. S. Schramm, L. Schunk, BOOST Camp 2018. https://indico.cern.ch/event/649482/contributions/3052141/attachments/1686853/2712969/BOOST_Camp_2018.pdf (2017)
75. ATLAS collaboration, G. Aad et al., The ATLAS experiment at the CERN large Hadron collider. J. Instrum. **3**, S08003 (2008)
76. ATLAS collaboration, G. Aad et al., Charged-particle multiplicities in pp interactions measured with the ATLAS detector at the LHC. New J. Phys. **13**, 053033 (2011). [1012.5104]
77. M. Cacciari, G.P. Salam, Dispelling the N^3 myth for the k_t jet-finder. Phys. Lett. **B641**, 57–61 (2006). [hep-ph/0512210]
78. J.M. Butterworth, J.P. Couchman, B.E. Cox, B.M. Waugh, KtJet: a C++ implementation of the K-perpendicular clustering algorithm. Comput. Phys. Commun. **153**, 85–96 (2003). [hep-ph/0210022]
79. M. Dasgupta, L. Magnea, G.P. Salam, Non-perturbative QCD effects in jets at hadron colliders. J. High Energy Phys. **0802**, 055 (2008). [0712.3014]
80. S. Catani, G. Turnock, B. Webber, L. Trentadue, Thrust distribution in e+ e- annihilation. Phys. Lett. **B263**, 491–497 (1991)
81. V.V. Sudakov, Vertex parts at very high-energies in quantum electrodynamics. Sov. Phys. J. Exp. Theor. Phys. **3**, 65–71 (1956)
82. B. Andersson, G. Gustafson, L. Lönnblad, U. Pettersson, Coherence effects in deep inelastic scattering. Z. Phys. **C43**, 625 (1989)
83. M. Dasgupta, G. Salam, Resummation of nonglobal QCD observables. Phys. Lett. **B512**, 323–330 (2001) . [hep-ph/0104277]
84. S. Catani, M. Ciafaloni, Many-gluon correlations and the quark form factor in QCD. Nucl. Phys. **B236**, 61 (1984)
85. Y.L. Dokshitzer, G. Marchesini, G. Oriani, Measuring color flows in hard processes: beyond leading order. Nucl. Phys. **B387**, 675–714 (1992)
86. E. Farhi, A QCD test for jets. Phys. Rev. Lett. **39**, 1587–1588 (1977)
87. M. Dasgupta, G.P. Salam, Accounting for coherence in interjet E(t) flow: a case study. J. High Energy Phys. **03**, 017 (2002). [hep-ph/0203009]
88. A. Bassetto, M. Ciafaloni, G. Marchesini, Jet structure and infrared sensitive quantities in perturbative QCD. Phys. Rept. **100**, 201–272 (1983)
89. M. Dasgupta, G.P. Salam, Resummed event shape variables in DIS. J. High Energy Phys. **08**, 032 (2002). [hep-ph/0208073]
90. A. Banfi, G. Corcella, M. Dasgupta, Angular ordering and parton showers for non-global QCD observables. J. High Energy Phys. **0703**, 050 (2007). [hep-ph/0612282]
91. A. Banfi, M. Dasgupta, Y. Delenda, Azimuthal decorrelations between QCD jets at all orders. Phys. Lett. **B665**, 86–91 (2008). [0804.3786]
92. A. Banfi, M. Dasgupta, K. Khelifa-Kerfa, S. Marzani, Non-global logarithms and jet algorithms in high-pT jet shapes. J. High Energy Phys. **1008**, 064 (2010). 1004.3483]
93. M. Dasgupta, K. Khelifa-Kerfa, S. Marzani, M. Spannowsky, On jet mass distributions in Z+jet and dijet processes at the LHC. J. High Energy Phys. **1210**, 126 (2012). [1207.1640]
94. A. Banfi, G. Marchesini, G. Smye, Away from jet energy flow. J. High Energy Phys. **0208**, 006 (2002) . [hep-ph/0206076]
95. G. Marchesini, A. Mueller, BFKL dynamics in jet evolution. Phys. Lett. **B575**, 37–44 (2003). [hep-ph/0308284]
96. I. Balitsky, Operator expansion for high-energy scattering. Nucl. Phys. **B463**, 99–160 (1996). [hep-ph/9509348]
97. Y.V. Kovchegov, Small × F(2) structure function of a nucleus including multiple pomeron exchanges, Phys. Rev. **D60**, 034008 (1999). [hep-ph/9901281]

98. E. Avsar, Y. Hatta and T. Matsuo, *Soft gluons away from jets: Distribution and correlation*, J. High Energy Phys. **0906**, 011 (2009). [0903.4285]
99. Y. Hatta, T. Ueda, Jet energy flow at the LHC. Phys. Rev. **D80**, 074018 (2009) . [0909.0056]
100. J. Jalilian-Marian, A. Kovner, A. Leonidov, H. Weigert, The Wilson renormalization group for low x physics: Towards the high density regime. Phys. Rev. **D59**, 014014 (1998). [hep-ph/9706377]
101. E. Iancu, A. Leonidov, L.D. McLerran, Nonlinear gluon evolution in the color glass condensate. 1. Nucl. Phys. **A692**, 583–645 (2001) . [hep-ph/0011241]
102. H. Weigert, Nonglobal jet evolution at finite N(c). Nucl. Phys. **B685**, 321–350 (2004). [hep-ph/0312050]
103. Y. Hatta, T. Ueda, Resummation of non-global logarithms at finite N_c. Nucl. Phys. **B874**, 808–820 (2013). [1304.6930]
104. S. Caron-Huot, Resummation of non-global logarithms and the BFKL equation. J. High Energy Phys. **03**, 036 (2018). [1501.03754]
105. R. Ángeles Martínez, M. De Angelis, J.R. Forshaw, S. Plätzer, M.H. Seymour, Soft gluon evolution and non-global logarithms. J. High Energy Phys. **05**, 044 (2018), [1802.08531]
106. J. Forshaw, J. Keates, S. Marzani, Jet vetoing at the LHC. J. High Energy Phys. **0907**, 023 (2009) . [0905.1350]
107. R.M.D. Delgado, J.R. Forshaw, S. Marzani, M.H. Seymour, The dijet cross section with a jet veto. J. High Energy Phys. **08**, 157 (2011). [1107.2084]
108. A.J. Larkoski, I. Moult, D. Neill, Non-global logarithms, factorization, and the soft substructure of jets. J. High Energy Phys. **09**, 143 (2015). [1501.04596]
109. A.J. Larkoski, I. Moult, D. Neill, the analytic structure of non-global logarithms: convergence of the dressed gluon expansion. J. High Energy Phys. **11**, 089 (2016). [1609.04011]
110. T. Becher, M. Neubert, L. Rothen, D.Y. Shao, Effective field theory for jet processes. Phys. Rev. Lett. **116**, 192001 (2016). [1508.06645]
111. T. Becher, M. Neubert, L. Rothen, D.Y. Shao, Factorization and resummation for jet processes. J. High Energy Phys. **11**, 019 (2016). [1605.02737]
112. M. Balsiger, T. Becher, D.Y. Shao, NLL' resummation of jet mass (2019). 1901.09038
113. C. Lee, G.F. Sterman, Universality of nonperturbative effects in event shapes. eConf C0601121, A001 (2006) . [hep-ph/0603066]
114. I.W. Stewart, F.J. Tackmann, W.J. Waalewijn, Dissecting soft radiation with factorization. Phys. Rev. Lett. **114**, 092001 (2015). [1405.6722]
115. Y.L. Dokshitzer, B. Webber, Power corrections to event shape distributions. Phys. Lett. **B404**, 321–327 (1997) . [hep-ph/9704298]
116. G. Salam, D. Wicke, Hadron masses and power corrections to event shapes. J. High Energy Phys. **0105**, 061 (2001) . [hep-ph/0102343]
117. S. Catani, M.L. Mangano, P. Nason, L. Trentadue, The resummation of soft gluon in hadronic collisions. Nucl. Phys. **B478**, 273–310 (1996) . [hep-ph/9604351]
118. N. Kidonakis, G. Oderda, G.F. Sterman, Evolution of color exchange in QCD hard scattering. Nucl. Phys. **B531**, 365–402 (1998). [hep-ph/9803241]
119. M. Diehl, J.R. Gaunt, Double parton scattering theory overview. Adv. Ser. Direct. High Energy Phys. **29**, 7–28 (2018). [1710.04408]
120. M. Dasgupta, A. Powling, L. Schunk, G. Soyez, Improved jet substructure methods: Y-splitter and variants with grooming. J. High Energy Phys. **12**, 079 (2016). [1609.07149]
121. G.P. Salam, L. Schunk, G. Soyez, Dichroic subjettiness ratios to distinguish colour flows in boosted boson tagging. J. High Energy Phys. **03**, 022 (2017). [1612.03917]
122. J.R. Andersen et al., Les Houches 2017: physics at TeV colliders Standard Model Working Group Report, in *10th Les Houches Workshop on Physics at TeV Colliders (PhysTeV 2017)*, Les Houches, June 5–23, 2017 (2018). 1803.07977
123. M. Dasgupta, A. Fregoso, S. Marzani, G.P. Salam, Towards an understanding of jet substructure. J. High Energy Phys. **1309**, 029 (2013). [1307.0007]
124. F.A. Dreyer, L. Necib, G. Soyez, J. Thaler, Recursive soft drop. J. High Energy Phys. **06**, 093 (2018). [1804.03657]

125. D. Krohn, J. Thaler, L.-T. Wang, Jet trimming. J. High Energy Phys. **1002**, 084 (2010) . [0912.1342]

126. S.D. Ellis, C.K. Vermilion, J.R. Walsh, Techniques for improved heavy particle searches with jet substructure. Phys.Rev. **D80**, 051501 (2009). [0903.5081]

127. D.E. Kaplan, K. Rehermann, M.D. Schwartz, B. Tweedie, Top tagging: a method for identifying boosted hadronically decaying top quarks. Phys. Rev. Lett. **101**, 142001 (2008). [0806.0848]

128. CMS collaboration, A Cambridge-Aachen (C-A) based Jet Algorithm for boosted top-jet tagging (2009). CMS-PAS-JME-09-001

129. CMS collaboration, *Boosted Top Jet Tagging at CMS*, CMS-PAS-JME-13-007.

130. M. Dasgupta, M. Guzzi, J. Rawling, G. Soyez, Top tagging : an analytical perspective. J. High Energy Phys. **09**, 170 (2018). [1807.04767]

131. A.J. Larkoski, J. Thaler, W.J. Waalewijn, Gaining (mutual) information about Quark/Gluon discrimination. J. High Energy Phys. **11**, 129 (2014). [1408.3122]

132. C.F. Berger, T. Kucs, G.F. Sterman, Event shape/energy flow correlations. Phys. Rev. **D68**, 014012 (2003). [hep-ph/0303051]

133. L.G. Almeida, S.J. Lee, G. Perez, G.F. Sterman, I. Sung, J. Virzi, Substructure of high-p_T jets at the LHC. Phys. Rev. **D79**, 074017 (2009). [0807.0234]

134. J. Gallicchio, M.D. Schwartz, Quark and Gluon Tagging at the LHC. Phys. Rev. Lett. **107**, 172001 (2011). [1106.3076]

135. J. Gallicchio, M.D. Schwartz, Quark and Gluon jet substructure. J. High Energy Phys. **04**, 090 (2013). [1211.7038]

136. J.R. Andersen et al., Les Houches 2015: Physics at TeV Colliders Standard Model Working Group Report, in *9th Les Houches Workshop on Physics at TeV Colliders (PhysTeV 2015)*, Les Houches, June 1–19, 2015 (2016). 1605.04692

137. F. Pandolfi, D. Del Re, Search for the standard model Higgs boson in the $H \to ZZ \to llqq$ decay channel at CMS. PhD thesis, Zurich, ETH, 2012

138. CMS collaboration, S. Chatrchyan et al., Search for a Higgs boson in the decay channel $H \to ZZ^{(*)} \to q\bar{q} l^- l^+$ in pp collisions at $\sqrt{s} = 7$ TeV. J. High Energy Phys. **04**, 036 (2012). [1202.1416]

139. B.T. Elder, J. Thaler, Aspects of track-assisted mass (2018). 1805.11109

140. J. Thaler, K. Van Tilburg, Identifying boosted objects with N-subjettiness. J. High Energy Phys. **03**, 015 (2011). [1011.2268]

141. I.W. Stewart, F.J. Tackmann, W.J. Waalewijn, N-jettiness: an inclusive event shape to veto jets. Phys. Rev. Lett. **105**, 092002 (2010). [1004.2489]

142. J. Thaler, K. Van Tilburg, Maximizing boosted top identification by minimizing N-subjettiness. J. High Energy Phys. **02**, 093 (2012). [1108.2701]

143. A.J. Larkoski, G.P. Salam, J. Thaler, Energy correlation functions for jet substructure. J. High Energy Phys. **1306**, 108 (2013). [1305.0007]

144. I. Moult, L. Necib, J. Thaler, New Angles on energy correlation functions. J. High Energy Phys. **12**, 153 (2016). [1609.07483]

145. A.J. Larkoski, I. Moult, D. Neill, Building a better boosted top tagger. Phys. Rev. **D91**, 034035 (2015). [1411.0665]

146. P.T. Komiske, E.M. Metodiev, J. Thaler, Energy flow polynomials: a complete linear basis for jet substructure. J. High Energy Phys. **04**, 013 (2018). [1712.07124]

147. P.T. Komiske, E.M. Metodiev, J. Thaler, Energy flow networks: deep sets for particle jets. J. High Energy Phys. **01**, 121 (2019). [1810.05165]

148. C. Frye, A.J. Larkoski, J. Thaler, K. Zhou, Casimir meets Poisson: improved quark/gluon discrimination with counting observables. J. High Energy Phys. **09**, 083 (2017). [1704.06266]

149. J. Thaler, L.-T. Wang, Strategies to identify boosted tops. J. High Energy Phys. **07**, 092 (2008). [0806.0023]

150. G. Soyez, G.P. Salam, J. Kim, S. Dutta, M. Cacciari, Pileup subtraction for jet shapes. Phys. Rev. Lett. **110**, 162001 (2013). [1211.2811]

151. M. Field, G. Gur-Ari, D. A. Kosower, L. Mannelli and G. Perez, *Three-Prong Distribution of Massive Narrow QCD Jets*, Phys. Rev. **D87** (2013) 094013, [1212.2106].
152. S.D. Ellis, A. Hornig, T.S. Roy, D. Krohn, M.D. Schwartz, Qjets: a non-deterministic approach to tree-based jet substructure. Phys. Rev. Lett. **108**, 182003 (2012). [1201.1914]
153. S.D. Ellis, A. Hornig, D. Krohn, T.S. Roy, On statistical aspects of Qjets. J. High Energy Phys. **01**, 022 (2015). [1409.6785]
154. CMS collaboration, A.M. Sirunyan et al., Search for low-mass resonances decaying into bottom quark-antiquark pairs in proton-proton collisions at $\sqrt{s} = 13$ TeV. Phys. Rev. (2018). [1810.11822]
155. J. Dolen, P. Harris, S. Marzani, S. Rappoccio, N. Tran, Thinking outside the ROCs: designing decorrelated taggers (DDT) for jet substructure. J. High Energy Phys. **05**, 156 (2016). [1603.00027]
156. I. Moult, B. Nachman and D. Neill, *Convolved Substructure: Analytically Decorrelating Jet Substructure Observables*, J. High Energy Phys. **05**, 002 (2018). [1710.06859]
157. D. Napoletano, G. Soyez, Computing N-subjettiness for boosted jets. J. High Energy Phys. **12**, 031 (2018). [1809.04602]
158. M. Dasgupta, A. Powling, A. Siodmok, On jet substructure methods for signal jets. J. High Energy Phys. **08**, 079 (2015). [1503.01088]
159. K. Kondo, Dynamical likelihood method for reconstruction of events with missing momentum. 1: method and toy models. J. Phys. Soc. Jpn. **57**, 4126–4140 (1988)
160. D0 collaboration, V.M. Abazov et al., A precision measurement of the mass of the top quark. Nature **429**, 638–642 (2004). [hep-ex/0406031]
161. P. Artoisenet, P. de Aquino, F. Maltoni, O. Mattelaer, Unravelling $t\bar{t}h$ via the matrix element method. Phys. Rev. Lett. **111**, 091802 (2013), [1304.6414]
162. K. Cranmer, T. Plehn, Maximum significance at the LHC and Higgs decays to muons. Eur. Phys. J. **C51**, 415–420 (2007). [hep-ph/0605268]
163. J.R. Andersen, C. Englert, M. Spannowsky, Extracting precise Higgs couplings by using the matrix element method. Phys. Rev. **D87**, 015019 (2013). [1211.3011]
164. D.E. Soper, M. Spannowsky, Finding physics signals with shower deconstruction. Phys. Rev. **D84**, 074002 (2011) [1102.3480]
165. D.E. Soper, M. Spannowsky, Finding top quarks with shower deconstruction. Phys. Rev. **D87**, 054012 (2013). [1211.3140]
166. ATLAS collaboration, Performance of shower deconstruction in ATLAS. ATLAS-CONF-2014-003
167. T. Plehn, G.P. Salam, M. Spannowsky, Fat jets for a light Higgs. Phys. Rev. Lett. **104**, 111801 (2010). [0910.5472]
168. T. Plehn, M. Spannowsky, M. Takeuchi, D. Zerwas, Stop reconstruction with tagged tops. J. High Energy Phys. **10**, 078 (2010). [1006.2833]
169. G. Kasieczka, T. Plehn, T. Schell, T. Strebler, G.P. Salam, Resonance searches with an updated top tagger. J. High Energy Phys. **06**, 203 (2015). [1503.05921]
170. Code for the cms top tagger in cms-sw. https://github.com/cms-sw/cmssw/blob/master/RecoJets/JetAlgorithms/interface/CMSTopTagger.h
171. Code for the heptoptagger. https://www.ippp.dur.ac.uk/~mspannow/heptoptagger.html
172. Code for shower deconstruction. https://www.ippp.dur.ac.uk/~mspannow/shower-deconstruction.html
173. M. Cacciari, G.P. Salam, G. Soyez, SoftKiller, a particle-level pileup removal method. Eur. Phys. J. **C75**, 59 (2015). [1407.0408]
174. D. Bertolini, P. Harris, M. Low, N. Tran, Pileup per particle identification. J. High Energy Phys. **10**, 059 (2014). [1407.6013]
175. M. Cacciari, G.P. Salam, Pileup subtraction using jet areas. Phys. Lett. **B659**, 119–126 (2008). [0707.1378]
176. SM MC Working Group, SM and NLO MULTILEG Working Group collaboration, J. Alcaraz Maestre et al., The SM and NLO multileg and SM MC working groups: summary

report, in *Proceedings, 7th Les Houches Workshop on Physics at TeV Colliders*, Les Houches, May 30–June 17, 2011 (2012), pp. 1–220. 1203.6803

177. S. Marzani, L. Schunk, G. Soyez, A study of jet mass distributions with grooming. J. High Energy Phys. **07**, 132 (2017). [1704.02210]

178. T. Sjöstrand, S. Ask, J.R. Christiansen, R. Corke, N. Desai, P. Ilten et al., An introduction to PYTHIA 8.2. Comput. Phys. Commun. **191**, 159–177 (2015). [1410.3012]

179. P. Skands, S. Carrazza, J. Rojo, Tuning PYTHIA 8.1: the Monash 2013 tune. Eur. Phys. J. **C74**, 3024 (2014). [1404.5630]

180. S. Marzani, L. Schunk, G. Soyez, The jet mass distribution after Soft Drop. Eur. Phys. J. **C78**, 96 (2018). [1712.05105]

181. P. Gras, S. Hoeche, D. Kar, A. Larkoski, L. Lönnblad, S. Platzer et al., Systematics of quark/gluon tagging. J. High Energy Phys. **07**, 091 (2017). [1704.03878]

182. S. Bright-Thonney, B. Nachman, Investigating the topology dependence of quark and gluon jets. 1810.05653

183. Z. Nagy, Next-to-leading order calculation of three-jet observables in hadron-hadron collisions. Phys. Rev. **D68**, 094002 (2003). [hep-ph/0307268]

184. J.M. Campbell, R.K. Ellis, Update on vector boson pair production at hadron colliders. Phys. Rev. **D60**, 113006 (1999). [hep-ph/9905386]

185. J.M. Campbell, R.K. Ellis, C. Williams, Vector boson pair production at the LHC. J. High Energy Phys. **07**, 018 (2011). [1105.0020]

186. J.M. Campbell, R.K. Ellis, W.T. Giele, A multi-threaded version of MCFM. Eur. Phys. J. **C75**, 246 (2015). [1503.06182]

187. A.J. Larkoski, I. Moult, D. Neill, Toward multi-differential cross sections: measuring two angularities on a single jet. J. High Energy Phys. **1409**, 046 (2014). [1401.4458]

188. A.J. Larkoski, I. Moult, D. Neill, Analytic boosted Boson discrimination. J. High Energy Phys. **05**, 117 (2016). [1507.03018]

189. M. Dasgupta, L. Schunk, G. Soyez, Jet shapes for boosted jet two-prong decays from first-principles. J. High Energy Phys. **04**, 166 (2016). [1512.00516]

190. A.J. Larkoski, I. Moult, D. Neill, Analytic boosted boson discrimination at the large Hadron collider. 1708.06760

191. S. Sapeta, Q.C. Zhang, The mass area of jets. J. High Energy Phys. **06**, 038 (2011). [1009.1143]

192. A. Larkoski, S. Marzani, J. Thaler, A. Tripathee, W. Xue, Exposing the QCD splitting function with CMS open data. Phys. Rev. Lett. **119**, 132003 (2017). [1704.05066]

193. CMS collaboration, Y. Chen, Jet substructure through splitting functions and mass in pp and PbPb collisions at 5.02 TeV with CMS. Nucl. Phys. **A967**, 512–515 (2017)

194. ALICE collaboration, D. Caffarri, Exploring jet substructure with jet shapes in ALICE. Nucl. Phys. **A967**, 528–531 (2017). [1704.05230]

195. K. Lapidus, M.H. Oliver, Hard substructure of quenched jets: a Monte Carlo study. 1711.00897

196. K.C. Zapp, Jet energy loss and equilibration. Nucl. Phys. **A967**, 81–88 (2017)

197. K. Tywoniuk, Y. Mehtar-Tani, Measuring medium-induced gluons via jet grooming. Nucl. Phys. **A967**, 520–523 (2017)

198. J. Casalderrey-Solana, Y. Mehtar-Tani, C.A. Salgado, K. Tywoniuk, Probing jet decoherence in heavy ion collisions. Nucl. Phys. **A967**, 564–567 (2017)

199. B. Nachman, M.L. Mangano, Observables for possible QGP signatures in central pp collisions. Eur. Phys. J. **C78**, 343 (2018). [1708.08369]

200. G.-Y. Qin, Modification of jet rate, shape and structure: model and phenomenology. Nucl. Part. Phys. Proc. **289-290**, 47–52 (2017)

201. G. Milhano, U.A. Wiedemann, K.C. Zapp, Sensitivity of jet substructure to jet-induced medium response. Phys. Lett. **B779**, 409–413 (2018). [1707.04142]

202. N.-B. Chang, S. Cao, G.-Y. Qin, Probing medium-induced jet splitting and energy loss in heavy-ion collisions. Phys. Lett. **B781**, 423–432 (2018). [1707.03767]

203. R. Kunnawalkam Elayavalli, K.C. Zapp, Medium response in JEWEL and its impact on jet shape observables in heavy ion collisions. J. High Energy Phys. **07**, 141 (2017). [1707.01539]
204. D. Neill, Non-global and clustering effects for groomed multi-prong jet shapes. 1808.04897
205. A. Tripathee, W. Xue, A. Larkoski, S. Marzani, J. Thaler, Jet substructure studies with CMS open data. Phys. Rev. **D96**, 074003 (2017). [1704.05842]
206. J. Gallicchio, M.D. Schwartz, Seeing in color: jet superstructure. Phys. Rev. Lett. **105**, 022001 (2010). [1001.5027]
207. ATLAS collaboration, M. Aaboud et al., Performance of top-quark and W-boson tagging with ATLAS in Run 2 of the LHC (2018). 1808.07858
208. ATLAS collaboration, G. Aad et al., Identification of high transverse momentum top quarks in pp collisions at $\sqrt{s} = 8$ TeV with the ATLAS detector. J. High Energy Phys. **06**, 093 (2016). [1603.03127]
209. ATLAS collaboration, G. Aad et al., Identification of boosted, hadronically decaying W bosons and comparisons with ATLAS data taken at $\sqrt{s} = 8$ TeV. Eur. Phys. J. **C76**, 154 (2016). [1510.05821]
210. ATLAS collaboration, Jet mass reconstruction with the ATLAS Detector in early Run 2 data. ATLAS-CONF-2016-035
211. A.J. Larkoski, I. Moult, D. Neill, Power counting to better jet observables. J. High Energy Phys. **12**, 009 (2014). [1409.6298]
212. G.C. Fox, S. Wolfram, Observables for the analysis of event shapes in e+ e− annihilation and other processes. Phys. Rev. Lett. **41**, 1581 (1978)
213. ATLAS collaboration, G. Aad et al., Measurement of kT splitting scales in W->lv events at sqrt(s)=7 TeV with the ATLAS detector. Eur. Phys. J. **C73**, 2432 (2013). [1302.1415]
214. L.G. Almeida, S.J. Lee, G. Perez, I. Sung, J. Virzi, Top jets at the LHC. Phys. Rev. **D79**, 074012 (2009). [0810.0934]
215. ATLAS collaboration, G. Aad et al., ATLAS measurements of the properties of jets for boosted particle searches. Phys. Rev. **D86**, 072006 (2012). [1206.5369]
216. C. Chen, New approach to identifying boosted hadronically-decaying particle using jet substructure in its center-of-mass frame. Phys. Rev. **D85**, 034007 (2012). [1112.2567]
217. G. Louppe, M. Kagan, K. Cranmer, Learning to pivot with adversarial networks. 1611.01046
218. C. Shimmin, P. Sadowski, P. Baldi, E. Weik, D. Whiteson, E. Goul et al., Decorrelated jet substructure tagging using adversarial neural networks. Phys. Rev. **D96**, 074034 (2017). [1703.03507]
219. C. Englert, P. Galler, P. Harris, M. Spannowsky, Machine learning uncertainties with adversarial neural networks. 1807.08763
220. CMS collaboration, Top tagging with new approaches. CMS-PAS-JME-15-002
221. CMS collaboration, V. Khachatryan et al., Identification techniques for highly boosted W bosons that decay into hadrons. J. High Energy Phys. **12**, 017 (2014). [1410.4227]
222. ATLAS collaboration, G. Aad et al., Light-quark and gluon jet discrimination in pp collisions at $\sqrt{s} = 7$ TeV with the ATLAS detector. Eur. Phys. J. **C74**, 3023 (2014). [1405.6583]
223. CMS collaboration, Jet algorithms performance in 13 TeV data. CMS-PAS-JME-16-003
224. ATLAS collaboration, Quark versus gluon jet tagging using charged particle multiplicity with the ATLAS detector. Technical report ATL-PHYS-PUB-2017-009, CERN, Geneva, May, 2017
225. ATLAS collaboration, G. Aad et al., Jet mass and substructure of inclusive jets in $\sqrt{s} = 7$ TeV pp collisions with the ATLAS experiment. J. High Energy Phys. **05**, 128 (2012). [1203.4606]
226. ATLAS collaboration, M. Aaboud et al., Measurement of the soft-drop jet mass in pp collisions at $\sqrt{s} = 13$ TeV with the ATLAS detector Phys. Rev. Lett. **121**, 092001 (2018). [1711.08341]
227. CMS collaboration, A.M. Sirunyan et al., Measurement of the jet mass in highly boosted tt̄ events from pp collisions at $\sqrt{s} = 8$ TeV. Eur. Phys. J. **C77**, 467 (2017) [1703.06330]
228. CMS collaboration, A.M. Sirunyan et al., Measurements of the differential jet cross section as a function of the jet mass in dijet events from proton-proton collisions at $\sqrt{s} = 13$ TeV. J. High Energy Phys. **11**, 113 (2018). [1807.05974]

229. ATLAS collaboration, G. Aad et al., Measurement of jet charge in dijet events from $\sqrt{s}$=8 TeV pp collisions with the ATLAS detector. Phys. Rev. **D93**, 052003 (2016). [1509.05190]
230. CMS collaboration, A.M. Sirunyan et al., Measurements of jet charge with dijet events in pp collisions at $\sqrt{s}$ = 8 TeV. J. High Energy Phys. **10**, 131 (2017). [1706.05868]
231. CMS collaboration, A.M. Sirunyan et al., Measurement of the splitting function in pp and Pb-Pb collisions at $\sqrt{s_{NN}}$ = 5.02 TeV. Phys. Rev. Lett. **120**, 142302 (2018). [1708.09429]
232. CMS collaboration, A.M. Sirunyan et al., Inclusive search for a highly boosted Higgs boson decaying to a bottom quark-antiquark pair. Phys. Rev. Lett. **120**, 071802 (2018). [1709.05543]
233. ATLAS collaboration, Search for boosted resonances decaying to two b-quarks and produced in association with a jet at $\sqrt{s}$ = 13 TeV with the ATLAS detector. ATLAS-CONF-2018-052
234. K. Joshi, A.D. Pilkington, M. Spannowsky, The dependency of boosted tagging algorithms on the event colour structure. Phys. Rev. **D86**, 114016 (2012). [1207.6066]
235. ATLAS collaboration, M. Aaboud et al., Search for heavy particles decaying into top-quark pairs using lepton-plus-jets events in proton-proton collisions at $\sqrt{s}$ = 13 TeV with the ATLAS detector. Eur. Phys. J. **C78**, 565 (2018). [1804.10823]
236. CMS collaboration, A.M. Sirunyan et al., Search for resonant $t\bar{t}$ production in proton-proton collisions at $\sqrt{s}$ = 13 TeV. J. High Energy Phys. (2018). [1810.05905]
237. ATLAS collaboration, M. Aaboud et al., Search for $W' \rightarrow tb$ decays in the hadronic final state using pp collisions at $\sqrt{s}$ = 13 TeV with the ATLAS detector. Phys. Lett. **B781**, 327–348 (2018). [1801.07893]
238. ATLAS collaboration, G. Aad et al., Search for Higgs boson pair production in the $b\bar{b}b\bar{b}$ final state from pp collisions at $\sqrt{s}$ = 8 TeVwith the ATLAS detector. Eur. Phys. J. **C75**, 412 (2015). [1506.00285]
239. CMS collaboration, A.M. Sirunyan et al., Search for a massive resonance decaying to a pair of Higgs bosons in the four b quark final state in proton-proton collisions at $\sqrt{s}$ = 13 TeV. Phys. Lett. **B781**, 244–269 (2018). [1710.04960]
240. CMS collaboration, A.M. Sirunyan et al., Search for production of Higgs boson pairs in the four b quark final state using large-area jets in proton-proton collisions at $\sqrt{s}$ = 13 TeV. J. High Energy Phys. **01**, 040 (2019). [1808.01473]
241. ATLAS collaboration, G. Aad et al., Search for high-mass diboson resonances with boson-tagged jets in proton-proton collisions at $\sqrt{s}$ = 8 TeV with the ATLAS detector. J. High Energy Phys. **12**, 055 (2015). [1506.00962]
242. CMS collaboration, V. Khachatryan et al., Search for massive resonances in dijet systems containing jets tagged as W or Z boson decays in pp collisions at $\sqrt{s}$ = 8 TeV. J. High Energy Phys. **08**, 173 (2014). [1405.1994]
243. K. Agashe, H. Davoudiasl, G. Perez, A. Soni, Warped gravitons at the LHC and beyond. Phys. Rev. **D76**, 036006 (2007). [hep-ph/0701186]
244. G.C. Branco, P.M. Ferreira, L. Lavoura, M.N. Rebelo, M. Sher, J.P. Silva, Theory and phenomenology of two-Higgs-doublet models. Phys. Rep. **516**, 1–102 (2012). [1106.0034]
245. H. Georgi, M. Machacek, Doubly charged Higgs bosons. Nucl. Phys. **B262**, 463–477 (1985)
246. C. Hackstein, M. Spannowsky, Boosting Higgs discovery: the forgotten channel. Phys. Rev. **D82**, 113012 (2010). [1008.2202]
247. C. Englert, C. Hackstein, M. Spannowsky, Measuring spin and CP from semi-hadronic ZZ decays using jet substructure. Phys. Rev. **D82**, 114024 (2010). [1010.0676]
248. ATLAS collaboration, Search for pair production of Higgs bosons in the $b\bar{b}b\bar{b}$ final state using proton-proton collisions at $\sqrt{s}$ = 13 TeV with the ATLAS detector. ATLAS-CONF-2016-049
249. F. Krauss, P. Petrov, M. Schoenherr, M. Spannowsky, Measuring collinear W emissions inside jets. Phys. Rev. **D89**, 114006 (2014). [1403.4788]
250. ATLAS collaboration, M. Aaboud et al., Search for diboson resonances with boson-tagged jets in pp collisions at $\sqrt{s}$ = 13 TeV with the ATLAS detector. Phys. Lett. **B777**, 91–113 (2018). [1708.04445]
251. CMS collaboration, A.M. Sirunyan et al., Search for massive resonances decaying into WW, WZ, ZZ, qW, and qZ with dijet final states at $\sqrt{s}$ = 13 TeV, Phys. Rev. **D97**, 072006 (2018). [1708.05379]

252. J.M. Butterworth, J.R. Ellis, A.R. Raklev, G.P. Salam, Discovering baryon-number violating neutralino decays at the LHC. Phys. Rev. Lett. **103**, 241803 (2009). [0906.0728]
253. CMS collaboration, A search for light pair-produced resonances decaying into at least four quarks. CMS-PAS-EXO-17-022
254. ATLAS collaboration, G. Aad et al., A search for top squarks with R-parity-violating decays to all-hadronic final states with the ATLAS detector in $\sqrt{s}$ = 8 TeV proton-proton collisions. J. High Energy Phys. **06**, 067 (2016). [1601.07453]
255. G. Brooijmans, P. Schichtel, M. Spannowsky, Cosmic ray air showers from sphalerons. Phys. Lett. **B761**, 213–218 (2016). [1602.00647]
256. Pierre Auger collaboration, A. Aab et al., Measurement of the average shape of longitudinal profiles of cosmic-ray air showers at the Pierre Auger Observatory. J. Cosmol. Astropart. Phys. (2018). [1811.04660]
257. G. Louppe, K. Cho, C. Becot, K. Cranmer, QCD-aware recursive neural networks for jet physics. J. High Energy Phys. **01**, 057 (2019). [1702.00748]
258. T. Sjöstrand, S. Mrenna, P. Skands, A brief introduction to PYTHIA 8.1. Comput. Phys. Commun. **178**, 852–867 (2008). [0710.3820]

Printed in the United States
By Bookmasters